AF360735

Filter-Feeding in Marine Invertebrates

Filter-Feeding in Marine Invertebrates

Editor

Hans Ulrik Riisgård

MDPI • Basel • Beijing • Wuhan • Barcelona • Belgrade • Manchester • Tokyo • Cluj • Tianjin

Editor
Hans Ulrik Riisgård
University of Southern Denmark
Denmark

Editorial Office
MDPI
St. Alban-Anlage 66
4052 Basel, Switzerland

ISBN 978-3-0365-5867-7 (Hbk)
ISBN 978-3-0365-5868-4 (PDF)

Contents

Preface to "Filter-Feeding in Marine Invertebrates"

Considering the dominant role of the phytoplankton in primary production in the sea, it is understandable that filter-feeding is widespread, and filter feeders (or suspension feeders) are found in almost all animal classes represented in the sea. Filter-feeding animals are necessary links between suspended food particles (phytoplankton, free-living bacteria, and other members of the microbial loop) and higher trophic levels in marine food webs. In addition to many holo- and mero-planktonic organisms, such as copepods and invertebrate larvae that graze on the phytoplankton and other food particles in the water column, many filter-feeding animals such as bivalves, polychaetes, ascidians, bryozoans, and sponges graze on the phytoplankton in the near-bottom water. Particularly in shallow coastal waters and fjords, dense populations of filter-feeders may exert a pronounced grazing impact, which may keep the water clear (but not clean) in eutrophicated areas. On the other hand, the dense populations of filter-feeding jellyfish in such areas may exert a pronounced predation impact on grazing zooplankton, resulting in a phytoplankton boom, and making the water green.

As it appears from the above description, filter-feeding in marine invertebrates is obviously a huge and important marine biological research area, which cannot be even approximately covered by the present six articles that were received before the deadline for manuscript submissions (15 August 2022). However, although these articles deal with a limited and rather random selection of both topics and filter-feeding species, they give an update of at least certain aspects of marine biological research. Thus, the present articles deal with many important topics, such as: filtration rates, energy budgets, growth rates, bioenergetic modeling, filter-pump design, particle-capture mechanisms, functional morphology, and hydrodynamics studied in sponges, jellyfish, mussels, and other filter-feeding marine invertebrates. This makes the Special Issue relevant for all marine biologists.

Hans Ulrik Riisgård
Editor

Journal of
Marine Science and Engineering

MDPI

Article

Superfluous Feeding and Growth of Jellyfish *Aurelia aurita*

Hans Ulrik Riisgård

Marine Biological Research Centre, Department of Biology, University of Southern Denmark, 5300 Kerteminde, Denmark; hur@biology.sdu.dk

Abstract: According to a recently presented bioenergetic model for the weight-specific growth rate of jellyfish, *Aurelia aurita*, fed brine shrimp, *Artemia salina*, the specific growth will remain high and constant at prey concentrations > 6 *Artemia* l^{-1}. The aim of the present study was to verify this statement by conducting controlled feeding and growth experiments on small jellyfish in tanks. It was found that prey organisms offered in concentrations of 25, 50, and 100 *Artemia* l^{-1} resulted in specific growth rates in fair agreement with the model-predicted rates. The high prey concentrations resulted in superfluous feeding and production of pseudofeces which indicated that not all captured prey organisms were ingested but instead entangled in mucus and dropped. The high prey concentrations did not influence the filtration rate of the jellyfish.

Keywords: weight-specific growth; bioenergetic growth model; filtration rate; pseudofeces

Citation: Riisgård, H.U. Superfluous Feeding and Growth of Jellyfish *Aurelia aurita*. *J. Mar. Sci. Eng.* **2022**, *10*, 1368. https://doi.org/10.3390/jmse10101368

Academic Editor: Azizur Rahman

Received: 2 September 2022
Accepted: 19 September 2022
Published: 25 September 2022

Publisher's Note: MDPI stays neutral with regard to jurisdictional claims in published maps and institutional affiliations.

1. Introduction

The common filter-feeding jellyfish *Aurelia aurita* occurs in many coastal ecosystems and can be very abundant and exert a considerable predatory impact on zooplankton [1–9]. *A. aurita* swims by means of umbrella pulsation and prey organisms are captured by tentacles on the bell rim during the recovery stroke [10]. *A. aurita* has a life cycle that includes a pelagic medusa and a benthic polyp stage. Medusae reproduce sexually, and females release planula larvae that settle and metamorphose into polyps that produce ephyrae that develop into medusae [11,12]. In temperate waters, an annual life cycle of *A. aurita* is typical [12,13]. Thus, in temperate Danish waters, ephyrae are released in spring resulting in a distinct cohort of medusae that reproduce sexually during summer, followed by loss of body mass ("degrowth") and disappearance of medusae in late autumn [14,15].

The population density and individual size of *Aurelia aurita* have over the years been investigated in the shallow semi-enclosed Danish cove Kertinge Nor [14–16]. In this cove, the numerous medusae are characterized by their small umbrella diameter. The population predation impact exerted by numerous small *A. aurita*, with estimated zooplankton half-lives of only about 1 to 3 d, indicates that shortage of prey controls the maximum umbrella size of typically 30 to 50 mm in Kertinge Nor [14,15], although in some years up to 60 to 70 mm [16] before subsequent degrowth.

In a recent study, [17] presented a bioenergetic model for the weight-specific growth rate of *Aurelia aurita* fed on 3-day-old brine shrimp *Artemia salina* as a reference prey organism. According to this model, the specific growth rate increases linearly with prey concentration in the range of 1 to 6 *Artemia* l^{-1} but remains high and constant at prey concentrations > 6 *Artemia* l^{-1}. The aim of the present study was to verify this last-mentioned statement by conducting controlled feeding and growth experiments in tanks with small food-limited jellyfish from Kertinge Nor exposed to various high prey concentrations resulting in superfluous feeding.

2. Materials and Methods

2.1. Collection of Jellyfish

Small (<65 mm umbrella diameter) jellyfish, *Aurelia aurita*, were collected on 3 June 2022 in Kerting Nor, Denmark, and brought to the nearby Marine Biological Research Centre for feeding and growth experiments.

2.2. Laboratory Feeding and Growth Experiments

Jellyfish were kept in tanks and continuously fed with 3-day-old *Artemia salina* nauplii obtained from cysts. Therefore, every day a new cohort of *A. salina* was started in air-mixed 3 l flasks. *A. salina* nauplii were transferred to a magnetic stirrer mixed 10 l stock culture in glass flask. By means of a peristaltic pump, the *Artemia* prey organisms were continuously dosed from the stock flask to the jellyfish growth tank and the same water volume (6.5 ml min^{-1}) was simultaneously taken out by another channel of the dosing pump. All experiments were conducted at 13 °C in a temperature-controlled aquarium hall and 20 psu as measured at the collecting site.

The wanted steady-state prey concentrations of 25, 50, and 100 *Artemia* l^{-1} in 3 parallel growth experiments (Exp #1, #2 and #3) were ensured by adjusting the number of *Artemia* added to the 12 l jellyfish tank to match the steadily increasing clearance rate of the growing jellyfish during the experimental period. Every day all jellyfish were carefully taken out and placed with the aboral side on a millimeter paper to measure their umbrella diameter. The concentration of prey organisms in each of the jellyfish growth tanks was measured every day by taking out samples for counting under a stereo microscope to adjust the number of prey organisms that had to be added to maintain the wanted mean concentration. The experiments were started on 7 June 2022 with 6 small jellyfish in each of the 3 tanks (50 cm diameter Breeding Air Kreisel, Schuran Seawater Equipment, www.schuran.com, accessed on 18 September 2022) with slowly air-driven circulating seawater to keep the jellyfish freely and undisturbed swimming. The feeding experiments ran for 17 d.

2.3. Equations

In a recent study, Ref. [17] set up the following bioenergetic model for weight-specific growth rate of *Aurelia aurita* fed on 3-day-old *Artemia salina*: $\mu = (n \times 0.07 - 0.08)W^{-0.2}$, where W (mg) is the jellyfish dry weight and n is the number of *Artemia* l^{-1} in the range of 1 to 6 *Artemia* l^{-1}. At prey concentrations > 6 *Artemia* l^{-1} the model conforms to:

$$\mu_{model} = (6 \times 0.07 - 0.08)W^{-0.2} = 0.34W^{-0.2}. \tag{1}$$

The aim of the present study was to verify if this simple growth model applies to superfluous feeding jellyfish.

The filtration rate (=clearance rate, F_{exp}) of jellyfish in the tanks was experimentally measured by the steady-state method which is based on the principle: [prey organisms removed by jellyfish ($F_{est} \times C_c$)] = [number of prey organisms dosed from stock-culture flask ($Fl \times C_p$) − prey washed out with outflowing seawater ($Fl \times C_c$)], so that:

$$F_{exp} = (Fl \times C_p - Fl \times C_c)/C_c, \tag{2}$$

where Fl = dosing pump rate, C_p = concentration of *Artemia* in stock-culture flask, C_c = concentration of *Artemia* in jellyfish tank.

The average size of *Aurelia aurita* during the growth period was estimated as:

$$W_{avg} = (W_0 \times W_t)^{1/2}, \tag{3}$$

where W_0 and W_t express the mean individual body dry weight of jellyfish at time t_0 and time t_t, respectively.

The following allometric equation was used to estimate dry weight (W, mg) from umbrella diameter (d, mm) of *Aurelia aurita* medusae (≥ 10 mm), [14]:

$$W = 1.73 \times 10^{-3} \times d^{2.82}. \tag{4}$$

The following equation for filtration rate (F_{model}, l d^{-1}) of *Aurelia aurita* fed on 3-day *Artemia* as a function of dry weight W (mg) was used in the bioenergetic model [17] and adapted from [7]:

$$F_{model} = 3.9W^{0.78}. \tag{5}$$

3. Results

The increase in the size of *Aurelia aurita* fed 25, 50, and 100 *Artemia* l^{-1} are shown in Figures 1–3 along with inserted exponential regression lines and their equations. The exponents of the regression equations that express the mean weight-specific growth rates (μ_{exp}) are shown in Table 1 along with the model-predicted growth rate (μ_{model}).

Table 1. *Aurelia aurita*. Experimental data and calculated parameters for feeding and growth experiments (Exp #1, #2, #3). C_c = mean $\pm$ s.d. concentration of *Artemia* prey organisms in the tank. Umbrella diameter on Day 0 = d_0, and on Day 17 = d_{17}. Estimated dry weight on Day 0 = W_0 and on Day 17 = W_{17} using Equation (4). W_{avg} = average size during the 17-day time interval estimated as $W_{avg} = (W_0 \times W_{17})^{1/2}$, cf. Equation (3). The predicted mean weight-specific growth rate (μ_{model}) was estimated using Equation (1). The experimentally determined specific growth (μ_{exp}) was determined as the b-exponent in the exponential regression equation for lines shown in Figures 1–3. Values for μ_{exp} leaving out the first 3 days are shown in brackets. F_{exp} = experimentally measured individual filtration rate using Equation (2). F_{model} = filtration rate estimated using Equation (5).

Exp	C_c (ind. l^{-1})	d_0 (mm)	d_{17} (mm)	W_0 (mg)	W_{17} (mg)	W_{avg} (mg)	μ_{model} (% d^{-1})	μ_{exp} (% d^{-1})	F_{exp} (l d^{-1})	F_{model} (l d^{-1})
#1	25 ± 10	48 ± 4	82 ± 8	95 ± 21	444 ± 137	170	12.1	10.1 (10.3)	343 ± 130	214
#2	49 ± 11	65 ± 4	101 ± 3	226 ± 41	778 ± 68	419	10.2	8.9 (10.6)	306 ± 100	433
#3	105 ± 70	61 ± 5	86 ± 11	191 ± 39	507 ± 191	331	10.7	5.4 (5.8)	362 ± 153	360

Because relatively low specific growth rates may be expected in the beginning of the feeding period due to mobilization of digestion processes in the previously starving jellyfish, and further, because possible initial growth in body thickness may take place before subsequent increase in umbrella diameter takes place, the weight-specific growth rates (μ_{exp}) have also been calculated leaving out the first 3 days and shown in brackets in Table 1. It is seen that these values are somewhat higher and in fair agreement with the model-predicted values (μ_{model}). Thus, it may be concluded that the simple bioenergetic model Equation (1) applies for small superfluous feeding *Aurelia aurita*.

The experimentally measured average filtration rates (=clearance rate of 3-day-old *Artemia*, F_{exp}) are in fair agreement with the estimated (F_{model}) using Equation (5) although pseudofeces (mucus entangled with prey organisms) accumulated at the bottom of the tanks (Figure 4). Superfluous feeding took place in all experiments and the amount of pseudofeces accumulated increased with increasing prey concentration, most pronounced in Exp #3. Therefore, the tanks had to be cleaned every 3 to 4 days.

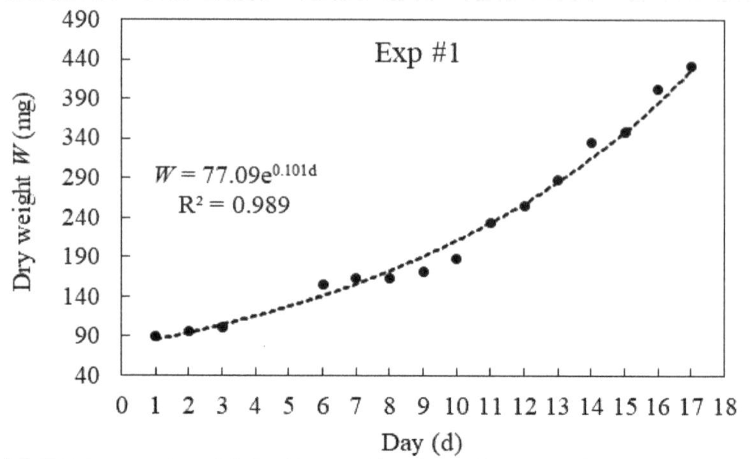

Figure 1. *Aurelia aurita.* Increase in dry weight of jellyfish fed 25 *Artemia* l^{-1}. The *b*-exponent of the exponential regression equations shows that the mean weight-specific growth rate is 10.1% d^{-1}.

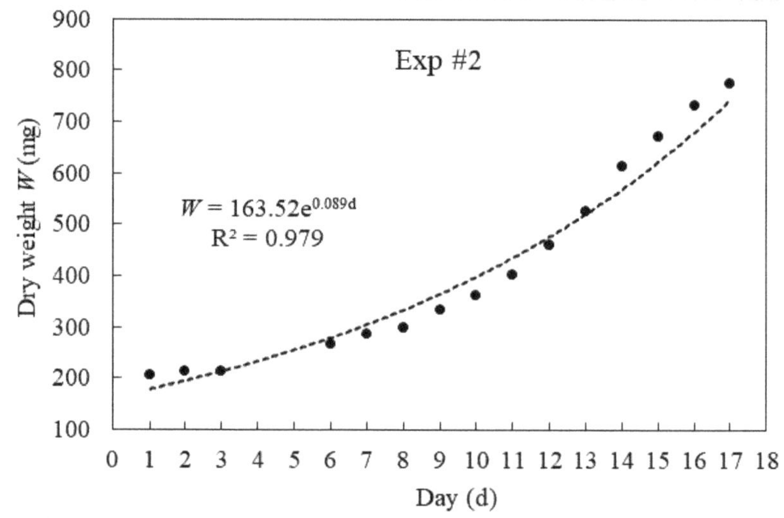

Figure 2. *Aurelia aurita.* Increase in dry weight of jellyfish fed 50 *Artemia* l^{-1}. The *b*-exponent of the exponential regression equations shows that the mean weight-specific growth rate is 8.9% d^{-1}.

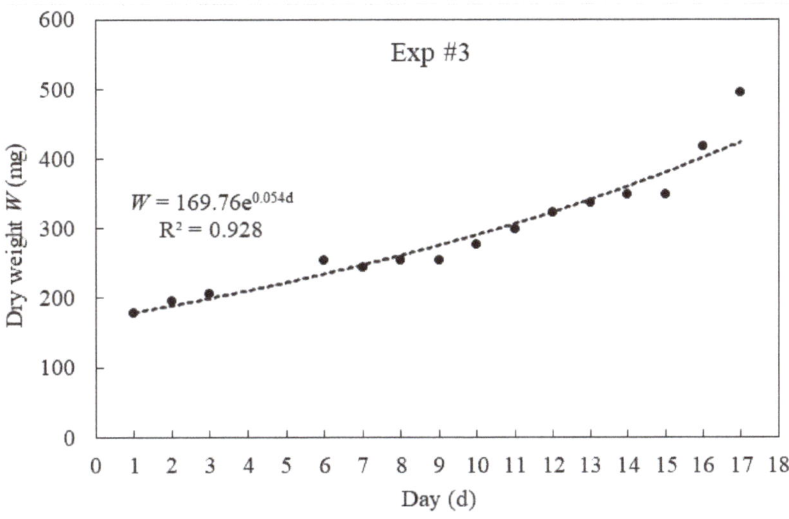

Figure 3. *Aurelia aurita*. Increase in dry weight of jellyfish fed 100 *Artemia* l^{-1}. The *b*-exponent of the exponential regression equations shows that the mean weight-specific growth rate is 5.4% d^{-1}.

Figure 4. (**A**) Experimental set-up with 3 Kreisler tanks on a bench and above that a shelf with peristaltic pumps dosing 3-day *Artemia* from well-mixed 10 l stock glass flasks placed on magnetic stirrers. The outflow water is collected in glass flasks below the tanks. (**B**) Tank with 6 small *Aurelia aurita*. (**C**) Accumulated pseudofeces on the bottom near a small glass funnel packed with cotton and connected to the opening of the outflow tube to prevent jellyfish from being sucked up. (**D**) Jellyfish pseudofeces consists of both living and dead 3-day *Artemia* (arrows) as well as black sphere-shaped unhatched cysts entangled in mucus.

4. Discussion

The growth potential of jellyfish in nature is generally not utilized due to a shortage of food. Thus, the increase in umbrella diameter of *Aurelia aurita* in the field was compared with "well-fed" jellyfish kept in tanks by [18]. The two groups showed near identical initial values at the start of May but began to diverge in June to become about 80 mm in the field and about 130 mm in the well-fed experiment. A re-plot of the well-fed *A. aurita* based on the estimated dry weight from umbrella diameter showed a systematic deviation between the data and regression curve because the weight-specific growth rate was not constant but decreased with a size that could be described by a power-function curve with $b = -0.24$ [17], which is close to the model-predicted b-value of -0.2 in Equation (1).

The present work shows that prey organisms offered in concentrations 4, 8, and 17 times above the lowest of six *Artemia* l^{-1}, which give rise to maximum growth of *Aurelia aurita* [17], results in superfluous feeding where a substantial number of the captured prey organisms are not being ingested but become entangled in mucus and dropped as pseudofeces. When the digestive system is filled up with prey the pseudofeces production ensures that the jellyfish may still utilize its growth potential while likewise, the filtration rate remains undisturbed up to at least 100 *Artemia* l^{-1} (Table 1). However, it should be stressed that *Artemia* used here as a reference prey organism is not among the natural zooplankton in the sea, but it is easily available food for cultured predatory organisms such as jellyfish. The 3-day-old *Artemia* have no escape behavior and are captured by jellyfish with higher efficiency than other prey. Thus, relative to *Artemia*, retention efficiency has been found to be 60% for rotifers, 35% for adult copepods, 22% for copepod nauplii, and 14% for mussel veligers [19].

The present study emphasizes that jellyfish are continuously filtering the ambient water at high rates and therefore in controlled feeding and growth experiments need to be fed continuously at relatively low prey concentrations. Thus, a mean individual filtration rate of 340 l d^{-1} or 236 mL min^{-1} as measured here in a 12 L tank with six jellyfish implies that the half-life of *Artemia* is: $t_{1/2} = 12,000/(236 \times 6) \times \ln2 = 5.9$ min. If all the prey organisms were offered one daily meal the prey organisms would have a mean residence time of only 5.9 min and would therefore rapidly be captured and most of them subsequently dropped to the bottom as pseudofeces resulting in suboptimal growth. As [14] noticed, when the guts of medusae were filled up with prey organisms at high prey concentrations, that part of the captured prey was killed and "apparently rejected instead of being digested". Obviously, a better understanding of the rejection and protection processes involved in superfluous feeding is needed, not least because knowledge of the actual prey ingestion rate and assimilation efficiency is important in bioenergetic studies on jellyfish.

5. Conclusions

When *Aurelia aurita* is offered prey concentrations 4, 8, and 17 times above the lowest needed for maximal growth this gives rise to superfluous feeding by which a substantial number of the captured prey organisms are not being ingested but become entangled in mucus and dropped as pseudofeces. Nevertheless, this does not influence the filtration rate of the jellyfish, which remains high and constant. Likewise, the weight-specific growth rate of *A. aurita* remains in fair agreement with the model-predicted growth. However, a better understanding of the processes involved in superfluous feeding in jellyfish is needed because knowledge of the actual prey ingestion rate is important in bioenergetic studies.

Funding: This research received no external funding.

Institutional Review Board Statement: Not applicable.

Informed Consent Statement: Not applicable.

Data Availability Statement: Not applicable.

Acknowledgments: Thanks are due to Marianne Croue and Théophile Magnac for technical assistance, to Josephine Goldstein for help with setting up the experiments, and to Poul S. Larsen for comments on the manuscript.

Conflicts of Interest: The author declares no conflict of interest.

References

1. Möller, H. Population dynamics of *Aurelia aurita* medusae in Kiel Bight, Germany (FRG). *Mar. Biol.* **1980**, *60*, 123–128. [CrossRef]
2. Båmstedt, U. Trophodynamics of scyphomedusae *Aurelia aurita*. Predation rate in relation to abundance, size and type of prey organism. *J. Plankton Res.* **1990**, *12*, 215–229. [CrossRef]
3. Behrends, G.; Schneider, G. Impact of *Aurelia aurita* medusae (Cnidaria, Scyphozoa) on the standing stock and community composition of mesozooplankton in the Kiel Bight (western Baltic Sea). *Mar. Ecol. Prog. Ser.* **1995**, *127*, 39–45. [CrossRef]
4. Lucas, C.H. Population dynamics of *Aurelia aurita* (Scyphozoa) from an isolated brackish lake, with particular reference to sexual reproduction. *J. Plankton Res.* **1996**, *18*, 987–1007. [CrossRef]
5. Uye, S.-I.; Fujii, N.; Takeoka, H. Unusual aggregations of the scyphomedusa *Aurelia aurita* in coastal waters along western Shikoku, Japan. *Plankton Biol. Ecol.* **2003**, *50*, 17–21.
6. Hansson, L.A.; Moeslund, O.; Kiørboe, T.; Riisgård, H.U. Clearance rates of jellyfish and their potential predation impact on zooplankton and fish larvae in a neritic ecosystem (Limfjorden, Denmark). *Mar. Ecol. Prog. Ser.* **2005**, *304*, 117–131. [CrossRef]
7. Møller, L.F.; Riisgård, H.U. Feeding, bioenergetics and growth in the common jellyfish *Aurelia aurita* and two hydromedusae, *Sarsia tubulosa* and *Aequorea vitrina*. *Mar. Ecol. Prog. Ser.* **2007**, *346*, 167–177. [CrossRef]
8. Møller, L.F.; Riisgård, H.U. Population dynamics, growth and predation impact of the common jellyfish *Aurelia aurita* and two hydromedusae, *Sarsia tubulosa* and *Aequorea vitrina* in Limfjorden (Denmark). *Mar. Ecol. Prog. Ser.* **2007**, *346*, 153–165. [CrossRef]
9. Matsakis, S.; Conover, R.J. Abundance and feeding of medusae and their potential impact as predators on other zooplankton in Bedford Basin (Nova Scotia, Canada) during spring. *Can. J. Fish. Aqua. Sci.* **2011**, *48*, 1419–1430. [CrossRef]
10. Costello, J.H.; Colin, S.P. Morphology, fluid motion and predation by the scyphomedusa *Aurelia aurita*. *Mar. Biol.* **1994**, *121*, 327–334. [CrossRef]
11. Hernroth, L.; Gröndahl, F. On the biology of *Aurelia aurita* (L.): 2. Major factors regulating the occurrence of ephyrae and young medusae in the Gullmar Fjord, western Sweden. *Bull. Mar. Sci.* **1985**, *37*, 567–576.
12. Lucas, C.H. Reproduction and life history strategies of the common jellyfish, *Aurelia aurita*, in relation to its ambient environment. *Hydrobiologia* **2001**, *451*, 229–246. [CrossRef]
13. Hamner, W.M.; Jenssen, R.M. Growth, degrowth, and irreversible cell differentiation in *Aurelia aurita*. *Am. Zool.* **1974**, *14*, 833–849. [CrossRef]
14. Olesen, N.J.; Frandsen, K.; Riisgård, H.U. Population dynamics, growth and energetics of jellyfish *Aurelia aurita* in a shallow fjord. *Mar. Ecol. Prog. Ser.* **1994**, *105*, 9–18. [CrossRef]
15. Goldstein, J.; Riisgård, H.U. Population dynamics and factors controlling degrowth of the common jellyfish, *Aurelia aurita*, in a temperate semi-enclosed cove (Kertinge Nor, Denmark). *Mar. Biol.* **2016**, *163*, 33–44. [CrossRef]
16. Lüskow, F.; Riisgård, H.U. Population predation impact of jellyfish (*Aurelia aurita*) controls the maximum umbrella size and somatic degrowth in temperate Danish waters (Kertinge Nor and Mariager Fjord). *Vie Milieu-Life Environ.* **2016**, *66*, 233–243.
17. Riisgård, H.U.; Larsen, P.S. Bioenergetic model and specific growth rates of jellyfish *Aurelia aurita*. *Mar. Ecol. Prog. Ser.* **2022**, *688*, 49–56. [CrossRef]
18. Ishii, H.; Båmstedt, U. Food regulation of growth and maturation in a natural population of *Aurelia aurita* (L.). *J. Plankton Res.* **1998**, *20*, 805–816. [CrossRef]
19. Riisgård, H.U.; Madsen, C.V. Clearance rates of ephyrae and small medusae of the common jellyfish *Aurelia aurita* offered different types of prey. *J. Sea Res.* **2011**, *65*, 51–57. [CrossRef]

Journal of
Marine Science and Engineering

MDPI

Article

High-Frequency Responses of the Blue Mussel (*Mytilus edulis*) Feeding and Ingestion Rates to Natural Diets

Laura Steeves [1,*], Antonio Agüera [2], Ramón Filgueira [2,3,*], Øivind Strand [2] and Tore Strohmeier [2]

[1]   Department of Biology, Dalhousie University, 1355 Oxford Street, Halifax, NS B3H 4R2, Canada
[2]   Institute of Marine Research, P.O. Box 1870, 5817 Bergen, Norway
[3]   Marine Affairs Program, Dalhousie University, 1355 Oxford Street, Halifax, NS B3H 4R2, Canada
*   Correspondence: laura.steeves@dal.ca (L.S.); ramon.filgueira@dal.ca (R.F.)

Abstract: The feeding activity of bivalves is understood to change in response to a suite of environmental conditions, including food quantity and quality. It has been hypothesized that, by varying feeding rates in response to the available diet, bivalves may be able to maintain relatively stable ingestion rates, allowing them to have constant energy uptake despite changes in food availability. The purpose of this study was to determine if the blue mussel *Mytilus edulis* responds to fluctuations in natural diets by changing feeding rates to maintain constant ingestion rates. Three four-day experiments were conducted to measure pumping and ingestion rates in response to natural fluctuations in food concentration (chlorophyll *a*). Experiments were conducted in a flow-through system over the spring season in south-western Norway. Pumping and ingestion rates were measured with high temporal resolution (every 20 min), which permitted the observation of the intra- and interindividual variability of feeding rates. Results show pumping rates varying within individuals over 4 days, and some individuals pumping on average at high rates (~5 Lh^{-1}), and some at low (~1 Lh^{-1}), despite being held in similar conditions. The pumping rate was generally not related to changes in food availability, and population-level ingestion rates increased with increasing food availability. These results suggest that, for this population of *M. edulis*, feeding rates may not vary with the available diet to produce constant ingestion over time.

Keywords: *Mytilus edulis*; pumping rate; ingestion rate; natural seston; filter-feeding; blue mussel

Citation: Steeves, L.; Agüera, A.; Filgueira, R.; Strand, Ø.; Strohmeier, T. High-Frequency Responses of the Blue Mussel (*Mytilus edulis*) Feeding and Ingestion Rates to Natural Diets. *J. Mar. Sci. Eng.* **2022**, *10*, 1290. https://doi.org/10.3390/jmse10091290

Academic Editor: Hans Ulrik Riisgård

Received: 15 August 2022
Accepted: 5 September 2022
Published: 13 September 2022

Publisher's Note: MDPI stays neutral with regard to jurisdictional claims in published maps and institutional affiliations.

1. Introduction

Suspension-feeding marine bivalves play important ecological roles by filtering plankton and detritus that are suspended in the water column and subsequently producing feces and pseudofeces that sink to the ocean floor. This top-down control on planktonic communities, as well as bottom-up control from bivalve excretion, can affect planktonic community structure and functioning [1–3]. Concomitantly, the quantity and quality of food (seston) available to suspension-feeding bivalves affects their performance in terms of growth and survival [4,5]. Many coastal marine environments are characterized by large fluctuations in seston composition and concentration, over both long (seasonal) and short (diel) timeframes [6]. Understanding the relationships between food availability and bivalve feeding behavior is crucial to predicting both bivalve growth and bivalve–ecosystem interactions.

Suspension-feeding bivalves have several mechanisms by which the quantity and composition of ingested food can be regulated. Pumping rate, the volume of water moved over the gills per unit time (PR), is a metric of feeding activity and may change by several liters per hour in an individual exposed to diets of differing concentration and composition [7–9]. Generally, the initiation of pumping is triggered when food concentration surpasses a minimum threshold level, which may vary both between species and populations [7,10–12]. As food levels continue to increase beyond the minimum threshold, *PR* may remain at a constant maximum or increase with food concentration [7,13–15]. When food levels become very high, *PR* may decline or become intermittent to avoid overloading the gills [16,17] or the digestive system [18,19], suggesting that the maximum ingestion rate (IR) has been reached. Bivalves may also regulate ingestion rates through the rejection of pseudofeces, a process which is usually not observed in low-seston environments ($< \sim 2.5$–5 mgL^{-1}), including the site used for this study [20]. Although bivalve *PR* in response to diets of varying composition has been extensively studied, a mechanistic understanding of this process is still relatively unknown [21,22].

For sessile species exposed to high levels of variation in the available diet, the ability to regulate the amount and quality of ingested food is an important mechanism in energy acquisition in bivalves. Although bivalves are exposed to frequently changing diets, these pre-ingestive mechanisms may help to maintain a relatively stable *IR* over time [23]. In the absence of pseudofeces production, *IR* may be estimated as a function of *PR* and food concentration [24]. For situations when food concentration is increasing and *PR* is decreasing, a relatively stable *IR* may be observed [25,26]. It has been theorized that this relationship between *PR* and food availability that can produce stable IRs may also contribute to constant energy uptake by bivalves in a fluctuating food environment [23]. In bivalves, the relationship between *IR* and food concentration is often modeled using Holling functional responses, which describe the relationship between prey density and predator consumption rates [27–29]. Holling functional responses may describe a linear increase (Type I) or asymptotic increases (Type II and III) in consumption rate with increasing prey density. The ability to accurately predict bivalve IRs in variable environmental conditions is a foundational step in predicting how bivalves acquire energy for growth.

The goal of this study was to examine relationships between *PR* and *IR* in response to fluctuation in natural diets and to explore the levels of intra- and interindividual variability in *PR* and IR. Often, the relationships between feeding, ingestion, and the food environment are studied using artificial diets (or natural seawater supplemented with artificial diets) in laboratory experiments [10,12,30]. However, experiments with natural diets are needed to understand the physiological responses of bivalves to the complexities of naturally occurring planktonic communities. Further, the current knowledge on the physiological responses in feeding activity to variability in diet comes primarily from environments with high seston concentration (>4 µgL^{-1}), either in laboratory studies or in sites where bivalves are cultivated [1,30]. However, many bivalves reside in environments that usually have lower seston concentrations (below the threshold for pseudofeces production), including the site used in this study [31,32]. It is important to study the physiology of bivalves in these low-seston environments to understand both the dynamics of natural populations and for potential future expansion of aquaculture farms due to space limitations in high-seston environments. Metrics of feeding and ingestion rates are often reported as an average of a group (e.g., one measurement on each individual) or by taking repeated measurements on the same individuals over the course of several hours [17,33]. These studies may overlook the short-term fluctuations in *PR* that can be captured with methodologies that allow high-frequency physiological measurements [34]. This study uses a novel methodology to estimate the feeding and ingestion rates of *M. edulis* with a high temporal resolution (every 18 min, for 4 days), using natural seawater under flow-through conditions. As seston concentration may change over the course of hours and days, this study aims to capture the functional feeding response of *M. edulis* over short timescales. *M. edulis* was selected as a model species as it is widely distributed and commercially important, and its feeding behavior has been extensively studied. It was hypothesized that, as the concentration and composition of the seston varied, *M. edulis* would vary *PR* to maintain constant IRs, above a minimum threshold of food concentration, following [23].

2. Material and Methods

2.1. Experimental Design

Three independent 4-day experiments were conducted to measure *Mytilus edulis* pumping rates (PR), ingestion rates (IR), and environmental conditions (Table 1). Dockside experiments were conducted in the spring of 2019 and 2020 at Austevoll Research Station (Institute of Marine Research), Norway (60°05′12.9″ N 5°15′51.5″ E). Experiments 1 and 2 (Exp. 1, 2) were conducted in May and June of 2019, respectively. Experiment 3 (Exp. 3) was conducted in April of 2020. Blue mussels (*M. edulis*) (30–60 mm) were collected from a local population and held at 3 m depth from a dock at the research station in hanging lantern nets for acclimation prior to all experiments. *M. edulis* were collected in February of 2019 (Exp. 1 and 2), and February 2020 (Exp. 3). All experiments used the same experimental set-up, in the same location. At least 24 h prior to each experiment, 10 experimental mussels were removed from the lantern, cleared of epibionts, and measured for shell length. Mussels were then placed in individual flow-through chambers (see [12] for chamber design). The individual chambers were designed to ensure the direct flow of water over the mussels and to avoid recirculation, preventing refiltration [35]. The size of the rectangular chambers (internal measurements) are as follows: width of 3.8 cm, length of 19.5 cm, and height of 8.1 cm. All chambers containing mussels were cleaned of feces every 12 h with a Pasteur pipette to avoid the resuspension of feces. Two chambers had water flowing through them with no mussels, to serve as controls.

Table 1. Summary of environmental and *M. edulis* physiology data from all experiments. Values represent the mean for environmental data and the median for physiological data. ±indicates standard deviation, and the coefficient of variation (%) is shown in parentheses.

	Exp. 1	Exp. 2	Exp. 3
Dates	May 07–11	June 04–08	April 06–13
Temperature (°C)	8.31 ± 0.16 (2)	10.51 ± 0.63 (6)	6.85 ± 0.14 (2)
Fluorescence (μg L^{-1})	0.67 ± 0.44 (66)	1.47 ± 0.47 (32)	2.99 ± 0.89 (30)
Suspended particulate matter (mg L^{-1})	1.68 ± 0.31 (18)	2.64 ± 0.52 (20)	1.92 ± 0.57 (30)
Energy (J L^{-1})	5.83 ± 1.74 (30)	11 ± 2.83 (26)	9.00 ± 1.87 (21)
Shell length (mm)	55.9 ± 1.6 (3)	59.5 ± 1.4 (2)	35.0 ± 2.5 (7)
Median pumping rate (L h^{-1})	2.0 ± 0.7 (35)	3.2 ± 0.4 (13)	3.1 ± 1.1 (35)
Median ingestion rate (μg h^{-1})	0.8 ± 1.2 (150)	4.4 ± 2.3 (52)	8.9 ± 4.1 (46)

Ambient, unfiltered seawater was pumped using an air pump (PlusAir: PA.15FVT) directly from the dock where mussels were being held to a water reservoir (600 L). From the water reservoir, seawater was gravity-fed to a header tank located directly above the individual flow-through chambers. From the header tank, water was flowed through 12 individual chambers. Following [36] flow-rates were regulated to aim for the 20–30% particle depletion of particles that are completely captured by mussels. The flow-rate through each chamber was measured a minimum of 4 times per day, and the flow-rates were corrected as needed through a regulating tap at the outflow.

2.2. Water Quality Measurements

Water temperature (°C) and fluorescence (as a proxy for chlorophyll *a*) (μgL^{-1}) measurements were taken every 30 min in the experimental water reservoir using a CTD (SAIV A/S Model 204). Water from the header tank was also filtered for suspended particulate matter (SPM; mgL^{-1}) and energy density (JL^{-1}). To do this, water filtered from a pressurized tank through pre-combusted and washed 90 mm filters (Whatman GF/D 2.7 μm pore width). The volumes filtered varied between 30–50 L, depending on the filtration rate. The timing of SPM and energy density measurements was similar for Exp. 1 and 2 and changed for Exp. 3 due to the availability of filters. For Exp. 1 and 2, water from the header tank was filtered for SPM and energy density measurements once every 12 h, with six replicates for each measurement. For Exp. 3, SPM and energy density were measured before and after the experiment (2 and 20 April 2020) in replicates of 10 and 5, respectively. All filters were rinsed twice with 50 mL of 0.5 M ammonium formate to remove any salts and kept frozen until analyzed. To measure SPM concentration, filters were dried in a 60 °C oven until weights were stable. Energy-density measurements were estimated from filters using a bomb calorimeter (BC, IKA model C6000) (Strohmeier et al. in prep). Filters were dried at 60 °C until stable weights were recorded, after which 500 mg of combustion aid (paraffin oil) was added to the filters to aid with complete combustion. Filters were combusted, and the measurement of temperature change (to the nearest 0.0001 K) was used to estimate energy density (JL^{-1}). Energy produced by the combustion aid and filter itself were subtracted from overall energy density to report the values of energy from the water column only.

2.3. Mussel Physiology

The feeding activity of *M. edulis* was measured as both *PR* and *IR* using the flow-through method [12,35,36]. This method relies on the accurate characterization of particles in the outflow of flow-through chambers (both from those containing a mussel and from the empty control chambers). In this experiment, the outflow of each chamber was connected to a normally closed solenoid valve. When a valve was closed, the outflow from that chamber would be directed to a drain. When opened, the outflow from that chamber was directed to an electronic laser particle counter (PAMAS S4031GO, GmbH, Hamburg, Germany), through silicone tubing. The solenoid valves from each individual chamber were opened sequentially, to ensure that the outflow from only one chamber at a time was delivered to the PAMAS. Solenoid valves were controlled by an Arduino Micro (3.X) connected to a relay board. The outflow of each chamber was sampled by the PAMAS for 60 s (volumetric equivalent of 10 mL), and then the particle counter was flushed for 30 s with the outflow of the following chamber before the next sample was recorded. This flushing period was employed to clean the PAMAS between samples. For Exp. 1 and 2, *PR* and *IR* measurements were taken on each individual and control every 18 min, and, for Exp. 3, measurements were taken on each individual every hour.

The PAMAS estimates particle size as equivalent spherical diameter (ESD, µm) and uses light scattering to count particles by size class at predefined intervals (0.5 µm in this study). From the estimates of particle counts for distinct size classes, both *PR* and *IR* were estimated. The pumping rate was estimated as:

$$PR = \left(\frac{P_c - P_b}{P_c} \right) \times FR \tag{1}$$

where *PR* is pumping rate (Lh^{-1}), P_c is the count of particles exiting the control chamber, P_b is the number of particles exiting the experimental chamber containing a bivalve, and *FR* is flow-rate through the chamber (Lh^{-1}) [37]. P_c and P_b were calculated using only particles understood to be completely captured on the gills (7.25, 7.75, and 8.25 µm ESD) [38]. Three size classes were used to minimize the potential error from a single particle-size count. Although larger particles (>8.25 µm ESD) are also expected to be completely captured, the abundance of these particles in the natural seston was low and were excluded to avoid introducing error into the calculation of *PR*. Chambers were monitored for pseudofeces production during all experiments, and none was observed at any time.

Pumping rates of individual mussels were standardized to gill area following [24]:

$$PR_{std} = PR \times \left(\frac{GA_{std}}{GA_{ind}} \right) \tag{2}$$

where PR_{std} is the standardized *PR*, GA_{std} is the gill area for the averaged size mussel from all experiments (46 mm, 22.38 cm^2), and GA_{ind} is the gill area for the individual mussel being standardized. The gill area was measured for all mussels in Exp. 1 and 2. Mussels were dissected by severing the anterior and posterior adductor muscles with a scalpel and separating both shell halves. In one half shell, gills were exposed by removing inner organs and mantle [39]. The gills were then floated in seawater to avoid contraction, and a photograph was taken from a top-down view. The area of one gill was then measured in ImageJ (v. 1.52 f) and multiplied by 8 (accounting for 4 gills, with 2 sides each), resulting in a total gill area of cm^2. For Exp. 3, no gill area pictures were taken, and gill area estimates were made from shell length following the relationship between length and gill area previously established for the same population of mussels: gill area [cm^2] = 0.0004 × length [mm]$^{2.85}$, r^2 = 0.79, n = 27 [24].

PR_{std} measurements were subsequently corrected for variations in temperature using an Arrhenius function [40]:

$$PR(T)_{std} = PR_1 \times \exp\left(\frac{T_A}{T_{AL}} - \frac{T_A}{T}\right) \times \frac{s(T)}{s(T_1)}$$
$$s(T) = \left(1 + \exp\left(\frac{T_{AL}}{T} - \frac{T_{AL}}{T_L}\right) + \exp\left(\frac{T_{AH}}{T_H} - \frac{T_{AH}}{T}\right)\right)^{-1} \tag{3}$$

where $PR(T)_{std}$ is the PR_{std} corrected to temperature T, T is the absolute temperature (281.15 K or 8 °C), T_1 is the reference temperature (K), PR_1 is the uncorrected PR at T_1, T_A is the Arrhenius temperature (5800 K), and T_{AL} (45430 K) and T_{AH} (31376 K) are the rates of PR decrease at the lower and upper temperature boundaries, respectively. T_L (275 K) and T_H (296 K) are the upper and lower temperature tolerance range, respectively. All Arrhenius parameters were obtained from van der Veer et al. (2006).

Ingestion rate was estimated from both PR and F values from the CTD as:

$$IR = PR(T)_{std} \times F \tag{4}$$

where IR is the ingestion rate (µgh^{-1}) calculated using PR standardized to both gill area and temperature, and F is chlorophyll a in µgL^{-1}. This calculation of IR is valid for conditions in which there is no production of pseudofeces.

2.4. Statistical Analyses

All statistical analyses were performed in R version 3.6.2 (RStudio version 1.4.1717). For periods during experiments wherein two control measurements were not reliably collected (e.g., if water was not sufficiently sampled from the outflow of the control chamber and air was introduced into the PAMAS, artificially reducing particle counts), all PR data were removed. If PR values for an individual mussel were unreasonably high (e.g., P_b counts ~0), it was assumed that no outflow water was being sampled by the PAMAS, and PR data for that individual was removed. For one sampling period (Exp. 1 and 2: 18 min, Exp. 3: 1 h), if fewer than 6 mussels were successfully sampled, all data were removed. Due to limitations in the precision of the particle counter, if $PR(T)_{std}$ was <0.2 Lh^{-1}, values were considered indistinguishable from 0, and the data were replaced with 0 but included in the data set. Within each experiment, a one-way repeated-measures ANOVA was run to test for differences in the PR between individual mussels. The PR data were checked for outliers and tested for normality using visual QQ-plots due to the large sample sizes within each dataset. The assumption of sphericity was checked with the Mauchly's test. If a significant effect of individual was observed on PR, post-hoc analyses with a Bonferroni adjustment was applied to observe pair-wise comparisons ($p < 0.0001$).

Within each experiment, median PR, IR, and chlorophyll a was visualized by fitting a locally estimated scatterplot smoothing (LOESS) regression [41]. For this regression, low-degree polynomials are fit to subsets of the data using weighted least squares. The size of the subsets of the data are determined using a smoothing parameter (α), which is a fraction of the number of datapoints. In this study, α = 0.1, resulting in low-degree polynomials being fit to the data every ~10 h. For the LOESS regression, PR, IR, and chlorophyll a datasets were interpolated with a linear function. To examine general relationships between population-level PR and chlorophyll a concentration, a non-linear function [12] was visualized on PR and chlorophyll a observations from all experiments combined:

$$PR = 5.35 - 0.67(F) + 0.56(ln(F) + 0.001/F) \tag{5}$$

where F is chlorophyll a concentration in µgL^{-1}. To observe how the data from these experiments may differ from those observed in [12], Equation (5) was fit to the data from these experiments, with new parameters being estimated with nonlinear least squares fitting (RStudio package: nlstools).

3. Results

3.1. Environmental Conditions

Environmental conditions varied between all experiments from April to June following a seasonal trend (Table 1). Mean temperature values ranged between 6.9 and 10.5 °C, with values being lowest in April (Exp. 3) and highest in June (Exp. 2). Mean chlorophyll *a* concentration varied from 0.7 (Exp. 1) to 3.0 μgL^{-1} (Exp. 3; Table 1). Suspended particulate matter (mgL^{-1}) and energy density (JL^{-1}) had similar trends with the lowest values in Exp. 1 (1.7 and 5.8, respectively) and the highest values in Exp. 2 (2.6 and 11.0, respectively; Table 1).

3.2. Pumping Rate

M. edulis in Exp. 1 had a median population-level *PR* (2.0 Lh^{-1}), with values varying over time between 0.1 and 3.6 Lh^{-1} (Table 1, Figure 1A). Notably, the population median *PR* was lowest between 9 and 10 May 2019 (Figure 1A). To further examine the variability in the population *PR*, examples of mussels with mussels high and low in interquartile range (IQR) in *PR* were analyzed (Figure 1B). At the same point in time, the *PR* between two mussels varied as much as ~3 Lh^{-1}, which was particularly noticeable at the end of the experiment (11 May 2019) (Figure 1B). Although both mussels periodically stopped pumping (*PR* = 0), the timing and frequency of closures varied between individuals (Figure 1B). Additionally, some individuals had relatively stable *PRs* compared to others (Figure 1C), with the coefficient of variation in *PR* ranging from 28 to 162%. Significant differences were observed between the *PR* of individual mussels in Exp. 1, with average *PRs* ranging from 0.8 to 2.8 Lh^{-1} (F (6, 168) = 143.7, $p < 0.0001$, generalized eta squared = 0.256). Post-hoc analyses with a Bonferroni adjustment revealed that all of the pairwise differences, between time points, were statistically significantly different ($p < 0.0001$, Figure 1C). Post-hoc analyses with a Bonferroni adjustment indicated a total of 6 statistically significant comparisons between mussel *PR* ($p < 0.0001$, Figure 1C).

M. edulis in Exp. 2 had a population-level median *PR* of 3.2 Lh^{-1}, and the population-level *PR* was also the most stable of all experiments, ranging between 1.2 and 4.0 Lh^{-1} (Table 1, Figure 2A). In Exp. 2, one individual was excluded from the population median *PR* calculation, as *PR* was often not distinguishable from zero (Figure 2C, indicated with an asterisk over the boxplot). In general, there was no extended period of time (e.g., days) over which the median population *PR* was generally higher or lower (Figure 2A). In examining the *PR* of the individuals with high and low IQR in *PR* (Figure 2B), it was observed that the individual with the low IQR had a highly stable *PR* over 4 days. This mussel pumped consistently at an intermediate rate of ~ 3 Lh^{-1}, with few interruptions, until the end of the experiment. Contrastingly, the individual with the high IQR showed generally high *PRs* for the first 3 days of the experiment (~5 Lh^{-1}) and low around the 4th day (~2 Lh^{-1}). This mussel abruptly stopped pumping several times during the first two days of the experiment for short periods of time, before returning to a relatively high *PR* (~4 Lh^{-1}) (Figure 2B). Towards the end of the experiment, this mussel had more gradual changes in *PR*, occurring over the course of several hours. Similar to Exp. 1, at a single point in time, there was, at times, a ~3 Lh^{-1} difference in *PR* between two individuals (Figure 2B). Variability in *PR* within individuals was generally lower than in Exp. 1, with a coefficient of variation in *PR* ranging from 11 to 91% (Figure 2C). Significant differences were observed in *PR* between individual mussels over time, with average *PRs* for each individual ranging from 2.0 to 3.7 Lh^{-1} (F(3.3, 1202) = 136.6, $p < 0.0001$, generalized eta squared = 0.227, Figure 2C). Post-hoc analyses with a Bonferroni adjustment indicated a total of 5 statistically significant comparisons between mussel *PR* ($p < 0.0001$, Figure 2C).

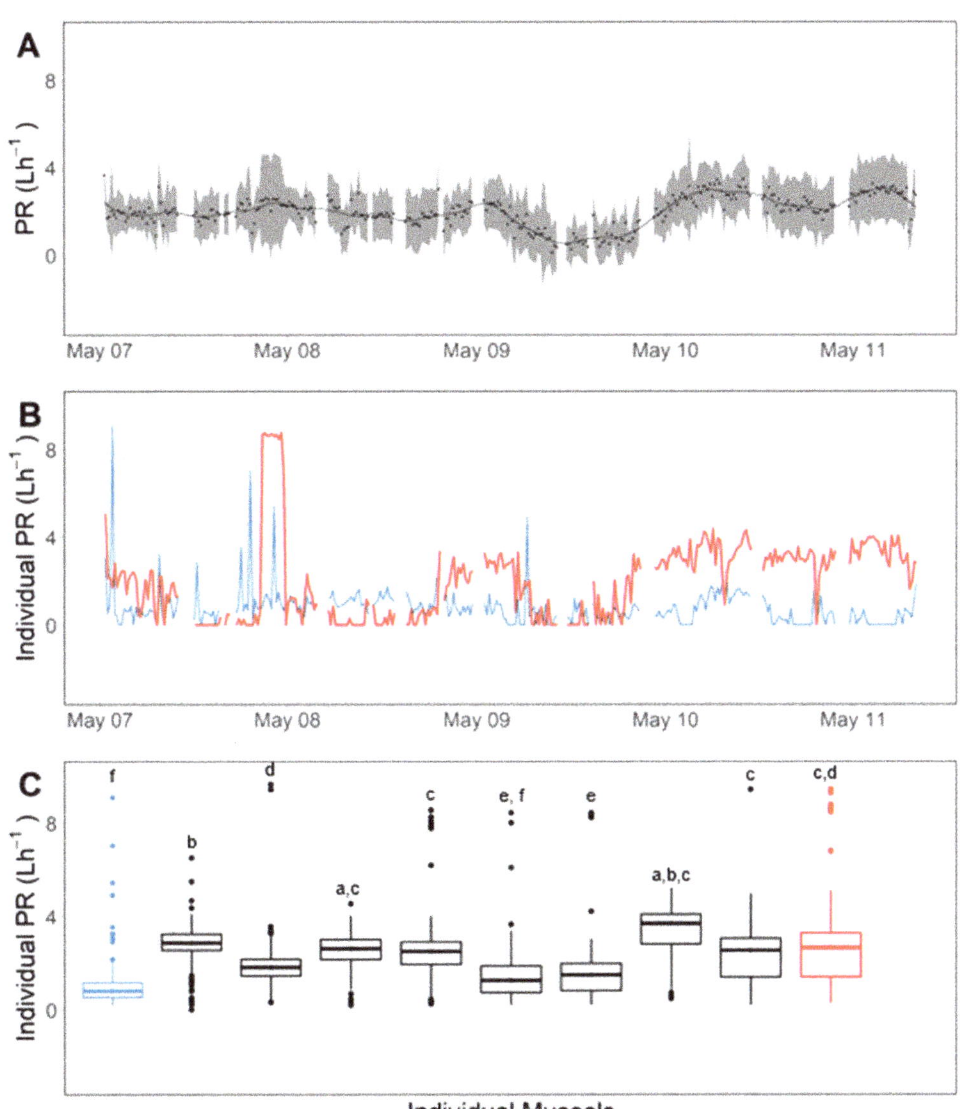

Figure 1. Summary of pumping rate (PR) (Lh^{-1}) data from Exp. 1: (**A**) Median *PR* of all individuals $\pm$ SD over 4 days. (**B**) Individual *PR* of two mussels with lowest (blue) and highest (red) interquartile range in PR. (**C**) Boxplots of summarized *PR* of all individuals over the entire experiment; letters a–f above boxplots indicate statistically significant differences at $p < 0.0001$.

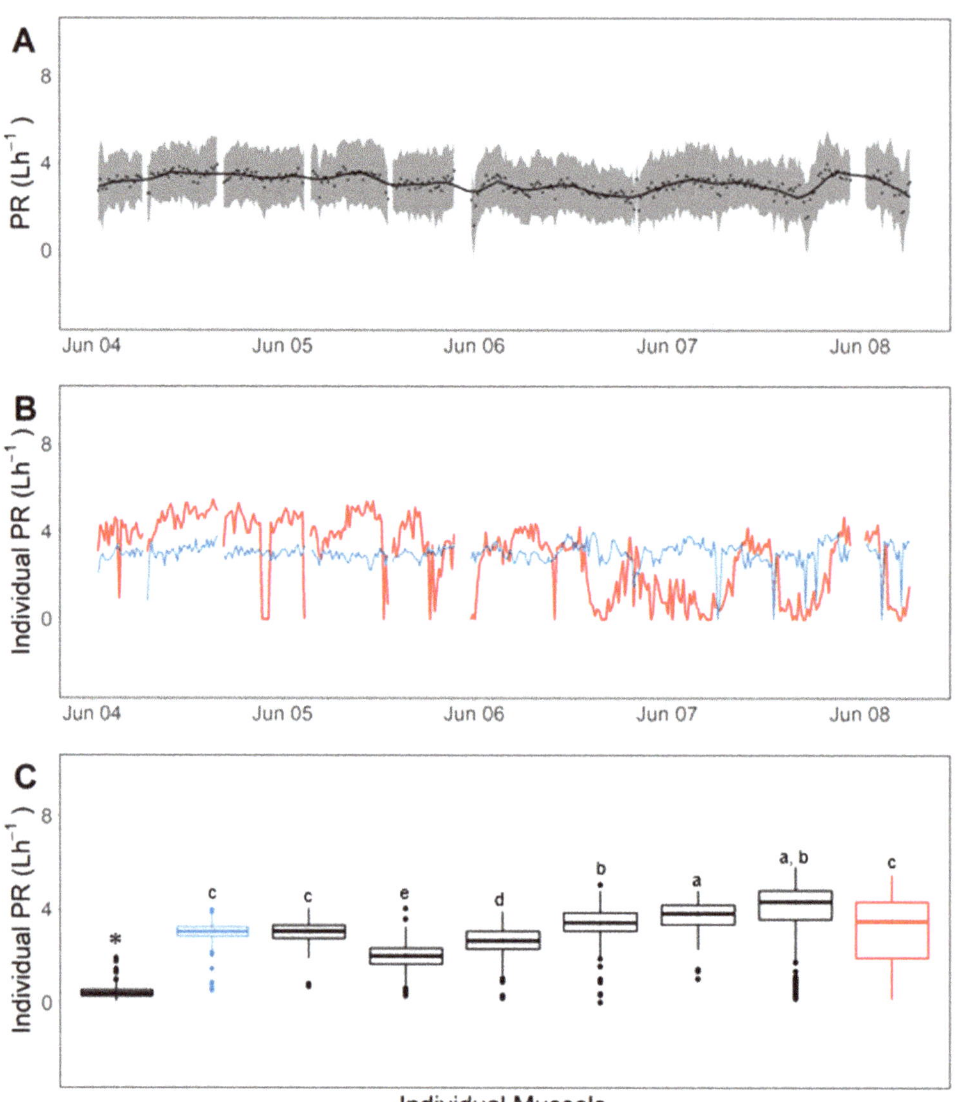

Figure 2. Summary of pumping rate (PR) (Lh^{-1}) data from Exp. 2: (**A**) median *PR* of all individuals $\pm$ SD over 4 days. (**B**) Individual *PR* of two mussels with low (blue) and high (red) interquartile range in PR. (**C**) Boxplots of the summarized *PR* of all individuals over the entire experiment; letters a–e above boxplots indicate statistically significant differences at $p < 0.0001$. * Indicates individual mussel not included in median measurements.

M. edulis in Exp. 3 had a median population-level *PR* of (3.1 Lh^{-1}); however, the variability in *PR* was markedly higher than in the first two experiments, both between and within individuals (Table 1, Figure 3A). The median population *PR* ranged from 1.0 to 7.5 Lh^{-1} (Figure 3A). Similar to in Exp. 2, there were no extended periods of high or low median population *PR*s, but *PR*s were generally variable over the 4 days. When examining the individuals with high and low IQR in *PR*, there was a marked difference between their *PR*s during the experiment. Although there were three mussels with lower IQR in *PR*

(Figure 3C), the fourth-lowest individual was selected to highlight in Figure 3B, as this individual had a more complete *PR* dataset during the experiment. The mussel with the low IQR in *PR* pumped at low rates over the course of the experiment (1.3 ± 0.9 Lh^{-1}), compared to the mussel with the highest IQR in *PR* (6.1 ± 2.4 Lh^{-1}) (Figure 3B, C). The high level of variability in the mussel pumping at 6.1 Lh^{-1} was driven by a decrease in *PR* over the last several days of the experiment (Figure 3B). Further, at a single point in time, there was a difference of ~7 Lh^{-1} in *PR* between two individuals (Figure 3B). The variability in *PR* within individuals was generally lower than Exp. 1, with a coefficient of variation in *PR* ranging from 31 to 135% (Figure 3C). Significant differences were observed in *PR* between individual mussels over time, with average *PR*s ranging from 0.5 to 6.1 Lh^{-1} ($F_{(3.7, 172)} = 91.2$, $p < 0.0001$, generalized eta squared = 0.64, Figure 3C). Post-hoc analyses with a Bonferroni adjustment indicated a total of 3 statistically significant comparisons between mussel *PR* ($p < 0.0001$, Figure 3C).

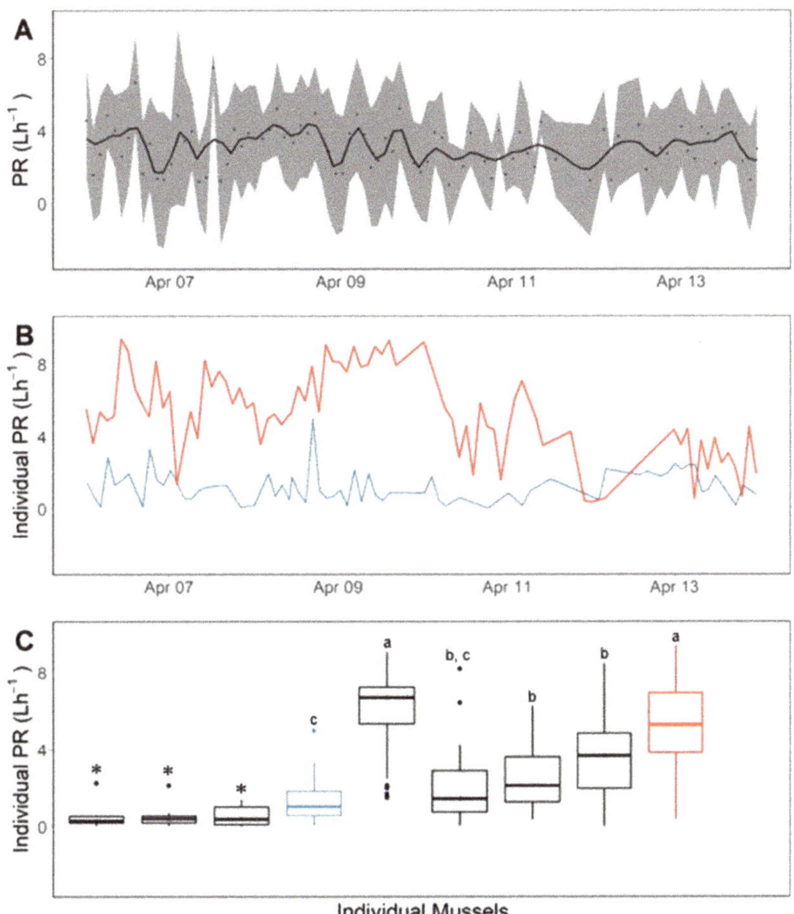

Figure 3. Summary of pumping rate (PR) (Lh^{-1}) data from Exp. 3: (**A**) median *PR* of all individuals $\pm$ SD over 4 days. (**B**) Individual *PR* of two mussels with low (blue) and high (red) interquartile range in PR. (**C**) Boxplots of the summarized *PR* of all individuals over the entire experiment; letters a–c above boxplots indicate statistically significant differences at $p < 0.0001$. * Indicates individual mussel not included in median measurements.

3.3. Ingestion Rate

In Exp. 1, the population-level median *IR* of *M. edulis* was 0.8 µgh^{-1} (Table 1, Figure 4A). The ingestion rate closely followed the pattern of median *PR* over time, with low rates between May 9 and 10 and rising on 11 May, matching the increase in *PR* (Figure 4A). The variability in population *IR* in Exp. 1 was 85%; however, the range was the lowest of all experiments (4.3 µgh^{-1}) (Table 1, Figure 4A). Exp. 2 has a population-level median *IR* of 4.4 µgh^{-1}, with a coefficient of variation of 36% and the second-largest range of all experiments (8.8 µgh^{-1}) (Table 1, Figure 4A). In Exp. 2, *IR* more closely followed the trend of chlorophyll *a* compared to *PR* over time, with a marked decrease in *IR* at the end of June 6 and an increase early on June 7, matching the pattern of chlorophyll *a* (Figure 4B). In Exp. 3, the population-level median *IR* was 8.9 µgh^{-1} with a coefficient of variation of 45% and the highest range of all experiments (17.4 µgh^{-1}) (Table 1, Figure 4C). Additionally, *IR* did not follow the pattern of either *PR* or chlorophyll *a* for the entire duration of the experiment (Figure 4C). Between April 8–9, *IR* followed the fluctuating pattern of *PR*; however, during the beginning and end of the experiment, *IR* followed the patterns in chlorophyll *a* (Figure 4C).

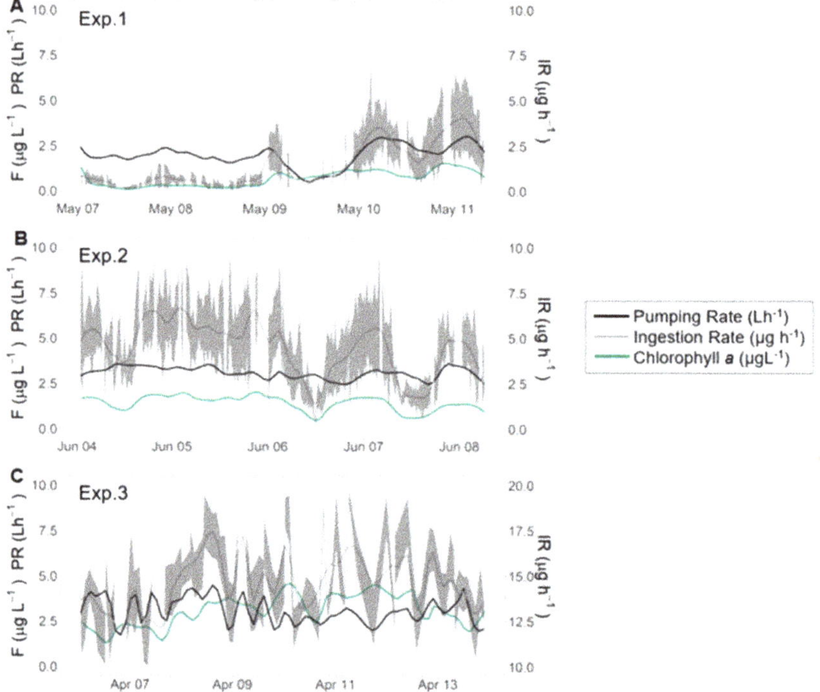

Figure 4. Pumping rate (PR) (Lh^{-1}) (black), chlorophyll *a* (F) (µgL^{-1}) (green), and ingestion rate (IR) (µgh^{-1}) (gray) averaged for all individual mussels in (**A**) Exp. 1, (**B**) Exp. 2. (**C**) Exp. 3. The gray shaded area is the standard deviation for IR.

3.4. Functional Responses to Food Availability

To examine the relationships between *PR*, *IR*, and food availability (chlorophyll *a*), the population-level results from all experiments were combined (Figure 5). When considering the population-level response in *PR* to chlorophyll *a* in all the experiments, no consistent trends were observed (Figure 5A). Additionally, the previously described relationship between *PR* and chlorophyll *a* in [12] did not well describe the relationship observed in this study (Figure 5A). *PR* generally did not increase with increasing chlorophyll *a*; however, interindividual variability in *PR* increased at higher chlorophyll *a* concentrations (>2 µgL^{-1}) (Figure 5A). For all experiments, population-level *IR* generally increased with increasing chlorophyll *a* (Figure 5B). At low concentrations of chlorophyll *a* (<2 µgL^{-1}), *IR* increases were highly linear with chlorophyll *a*; however, as the chlorophyll *a* concentration increased beyond 2 µgL^{-1}, the increase in *IR* became less linear (Figure 5B). Further, interindividual variability in *IR* increased in each subsequent experiment with increasing concentrations of chlorophyll *a* (particularly when chlorophyll *a* was >2 µgL^{-1}) (Figure 5B). The relationship between *IR* and increasing chlorophyll *a* was visualized with Holling functional responses (Type I, II, and III) to illustrate the potential response in *IR* being either linear or asymptotic.

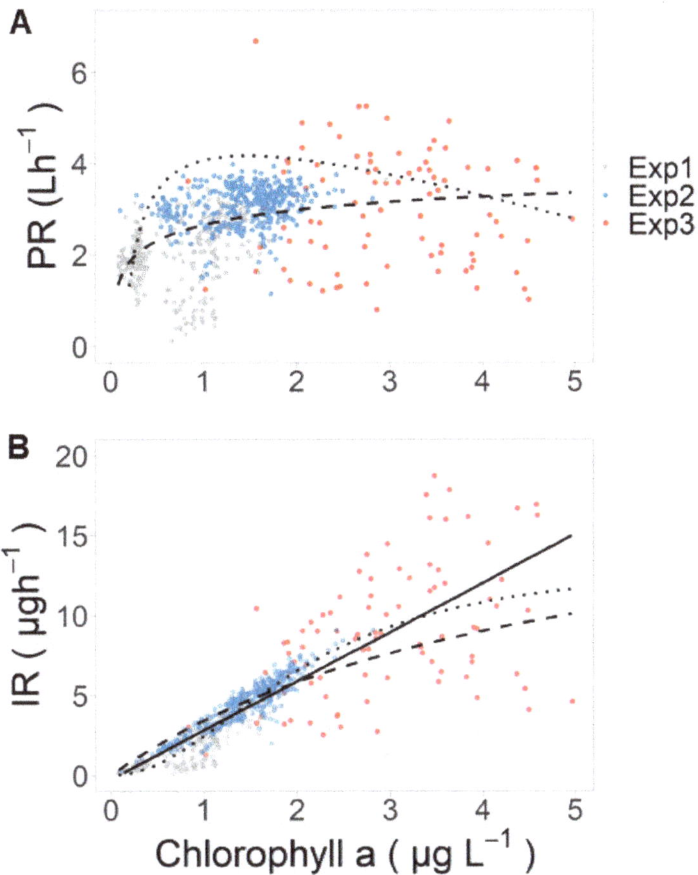

Figure 5. Relationships between (**A**) pumping rate (PR) (Lh^{-1}) and (**B**) ingestion rate (IR) (µgh^{-1}) and chlorophyll *a* (µgL^{-1}) for all experiments. The dotted line on (**A**) is drawn from Equation (5), and the dashed line is fitted from Equation (5) with parameters fit to this dataset: $PR = 2.69 - 0.02(F) + 0.53(ln(F) + 0.001/F)$. The drawn lines on (**B**) represent Holling functional responses (Type I, II, and III, solid, dashed, and dotted, respectively) to indicate the potential relationships between fluorescence and IR.

4. Discussion

This study used a novel flow-through methodology to measure feeding (pumping and ingestion rates) in *M. edulis* in response to natural fluctuations in diet. Although it has previously been hypothesized that bivalves alter pumping rates to maintain relatively constant ingestion rates [23], these compensatory processes were not observed in this study. Pumping rates displayed no consistent response to changes in food availability, as measured by chlorophyll *a* concentration. Further, *IR* generally increased with increasing food availability. The high frequency of pumping and ingestion rate measurement taken in this study permitted the exploration of both intra- and interindividual variability on a much finer temporal scale (minutes) compared to previous studies (hours/days/weeks). High levels of variability in pumping and ingestion rates were observed both between and within individuals during these 4-day experiments.

4.1. Feeding Activity in Response to Natural Fluctuations in Diet

The range of *PRs* recorded in this experiment (mean $\pm$ standard deviation: $3.0 \pm 1.8\,\mathrm{Lh^{-1}}$) are similar to those reported for *M. edulis* in similar environmental conditions [12,24,42]. Food concentration (or diet quantity) was characterized as chlorophyll *a* concentration and increased with each subsequent experiment from ~1 to ~3 $\mu\mathrm{gL^{-1}}$, which is within the range of values commonly reported during spring in this region [31,32]. In all experiments, *PR* generally was not related to changes in food concentration. Food concentration is understood to be a primary determinant of feeding rates in bivalves, wherein feeding is initiated at a minimum food concentration and quickly increases or switches to a maximum rate as food concentration increases [10,22]; finally, at food levels above a saturation threshold, feeding rates often decline, to avoid overloading the gills or because maximum *IR* may have been reached [16,43]. Although a cessation in the *PR* of mussels has been observed at low food concentrations (<0.5 $\mu\mathrm{gL^{-1}}$, [44], a previous study on the same population of *M. edulis* used in this study observed *PRs* between 2.5–4.7 $\mathrm{Lh^{-1}}$ at very low chlorophyll *a* concentrations (0.1–06 $\mu\mathrm{gL^{-1}}$) [12]. Further, a decline in *PR* was not expected, as food concentrations (<3 $\mu\mathrm{gL^{-1}}$) did not reach the saturation threshold expected to trigger a reduction in feeding rates [22,43]. Therefore, the lack of a relationship between population-level *PR* and chlorophyll *a* in any of the 4-day experiments is not unexpected for the low levels of chlorophyll *a* observed in this study.

In this population of *M. edulis*, relatively stable *PRs* have also been observed, despite changes in a diet of similar quantities (chlorophyll *a* concentration) [12]. It is possible that the lack of relationship between *PR* and chlorophyll *a* observed in this experiment indicates that, for individuals adapted to maximize ingestion rates in low-seston environments, *PR* is initiated at a very low food concentration and remains high as food concentration increases. Bivalves inhabiting low-seston environments have often been observed to have very high feeding rates in field studies [12,45–47]. At chlorophyll *a* levels much higher than those observed in this study (>3 $\mu\mathrm{gL^{-1}}$), the *PR* of *M. edulis* may decline; however, these conditions are not frequent in this region [31,32]. It has previously been recognized by [48] that the strategy of bivalves to regulate the amount of ingested material may vary by species, wherein *M. edulis* has often been observed to regulate ingestion through pseudofeces production, while continuing to pump at high rates [7,11,45]. As the range in diet observed in this study remained under the threshold for the production of pseudofeces, it is likely that the mussels were continuing to pump at high rates. The lack of a relationship between *PR* and chlorophyll *a* levels observed may also indicate that, for short-term fluctations in diet quantity, a physiological response in *PR* may not be elicited. This "time-averaged" behavior may be an explanation for why *PRs* do not change in response to diet changes that only last on the scale of minutes to hours [48].

Aspects of diet composition (or diet quality) that may affect feeding rates include seston load and the fraction of non-digestible inorganic material [9,17,19,29,49–51]. By characterizing the diet using chlorophyll *a*, some qualitative aspects of the diet known to influence *PR* may not be captured [9,17]. Although chlorophyll *a* increased from Exp. 1 to Exp. 3, the highest concentrations of suspended particulate matter and energy were observed in Exp. 2, indicating that diet quality was also changing between experiments. Although fluorescence concentration does not comprehensively describe the available diet, it is easily measured with high temporal frequency, compared to the more time-intensive methods required for the filtration of water for SPM and energy concentration [34].

Resultingly, high-temporal-resolution measurements of chlorophyll *a* concentration may provide one of the best available methods to take measurements of diet and feeding physiology on similar temporal scales.

The functional response of *IR* to food concentration in bivalves has been previously described using different Holling functional responses. Most commonly used are the Type II and III functional responses, which are characterized by stable IRs at high food concentrations [28,29]. The population-level *IR* in this experiment generally increased with increasing chlorophyll *a* concentration; however, this relationship had the highest slope when the food concentration was low (<2 μgL^{-1}). The population-level response in *IR* to increasing food concentrations in this study suggests that any of the Holling functional responses may statistically represent the observed relationship. However, the data collected in this study is heavily concentrated with observations at low food concentrations (<2 μgL^{-1}), compared to higher concentrations (~2–5 μgL^{-1}), which limits the ability to estimate an asymptotic relationship. Although a stable *IR* at high food levels has been previously hypothesized (Holling Type II and III) [23,25,26,52], it is likely that food levels in this experiment did not reach high enough concentrations to observe maximum and constant ingestion rates. As previously described, it is possible that the strategy of individuals adapted to low-seston environments may be to continuously pump at a high rate, resulting in increasing ingestion rates with increasing food concentration [12].

Despite the lack of the clear stabilization of ingestion rates at high food concentrations, the observations revealed increasing levels of inter-individual variability in both ingestion and pumping rates at high chlorophyll *a* concentrations. This variability in feeding physiology at increasing food concentrations may indicate the periodic stopping or slowing of feeding driven by digestive processes (e.g., gut capacity being reached, maximum *IR* being reached) [18,27,53,54]. Accordingly, it is possible that an asymptote in ingestion rates reflective of a Holling Type II or III response may emerge at higher food concentrations (e.g., >3 μgL^{-1}) if the periodic slowing or stopping of *PR* becomes more frequent at the population level.

4.2. Intra- and Interindividual Variability in Feeding Activity

The high temporal resolution of the methodology used in this experiment was selected to be able to examine both intra- and interindividual variability in pumping and ingestion rates in response to real-time fluctuations in diet. By observing the range of physiological rates within an individual over the scale of hours and days, it is possible to more accurately observe short-term fluctuations in feeding physiology in response to environmental variability in terms of food quantity and quality [55,56]. In previous studies, when physiological rates have been measured only one time per individual or repeatedly on an individual with coarse temporal resolution, it is possible to overlook the full range of intra- and interindividual variability over short timescales [34].

Inter-individual variability was observed during each 4-day experiment between the *PR*s of individual mussels. Despite being exposed to the same conditions, the average *PR* of the mussels ranged ~3 Lh^{-1} between individuals. Inter-individual variability in physiological rates, including feeding rates, has been explored as a potential explanation for different growth rates between fast- and slow-growing individuals [57], and similar inter-individual variability in feeding rates of bivalves exposed to the same conditions has been observed in other studies [58,59]. In this experiment, differences between experimental individuals were minimized by selecting *M. edulis* of the same size and age-class from the same location. The goal in selecting similar individuals was to minimize differences in inter-individual variability driven by factors not examined in this study. However, it is possible that there were differences between the *M. edulis* used in this study that were not accounted for, including sex (potentially influencing energetic requirements), genetic differences, and maternal effects [60–63]. Future experiments may consider further minimizing differences between individuals by rearing first-generation offspring together in common conditions (e.g., [64]) or by increasing the duration of the experiments to observe if average physiological rates between individuals are similar across longer periods of time (e.g., seasonally or annually).

Intraindividual variability was observed in all experiments, wherein *PR* and *IR* varied within individuals over the 4-day periods. Variability in the feeding physiology of bivalves may be driven by changes in environmental conditions, including those previously discussed (e.g., temperature, diet) [14,15,65]. However, the periodic cessation of feeding in *M. edulis* observed in this study was unsynchronized between individuals, suggesting that feeding rates may have been regulated by internal drivers rather than external environmental conditions under the environmental conditions of these experiments. For example, if gut capacity is reached, feeding rates may slow down; however, gut capacity may not be reached at the same time for all individuals [18,19]. The high temporal resolution of the *PR* data presented here indicates that *PR* activity varies between individuals in terms of how consistent *PR* is over time, maximum and minimum rates, and how quickly *PR* may increase or decrease (e.g., on the scale of minutes to hours). Observing intraindividual variability in the feeding physiology of *M. edulis* and characteristics of the natural diet at high temporal resolution provides insights into the drivers of the feeding physiology of bivalves. Further, although the unsynchronized individual responses observed in these experiments suggest that *PR* is not driven by environmental factors, their influence in feeding physiology cannot be disregarded, and further experiments under broader environmental conditions are warranted.

4.3. Energy Acquisition

Chlorophyll *a* is used in this study as a proxy for food concentration; however, chlorophyll *a* is limited as a proxy for the amount of food that is captured and ingested from the seston by *M. edulis*. Chlorophyll *a* alone is not able to capture the complexity of the seston in terms of particle sizes and surface properties, which both may affect particle capture efficiency [66]. Capture efficiency describes the proportion of particles captured on the gill, compared to those in the water, and is often described according to particle size, wherein capture efficiency increases with increasing particle size to some maximum, beyond which all particles are captured [67,68]. However, capture efficiency has also been related to other particle characteristics including hydrophobicity [69], lectin–carbohydrate interactions [70], and chlorophyll *a* [71]. Additionally, capture efficiency has been observed to vary in *M. edulis* across seasons in response to natural seston composition [24,72]. As *IR* is described in this experiment using chlorophyll *a*, if changes in capture efficiency occurred, it would not be accounted for in estimates of ingestion. Further, estimation of *IR* using chlorophyll *a* instead of the total volume of ingested material may not be used to estimate gut capacity, which may limit maximum ingestion rates [18,19].

It has been theorized that, as the quality and quantity of their diet changes, bivalves will make use of behavioral and physiological mechanisms to maintain constant energy uptake [10,18,23,26]. Although in this study, constant ingestion rates were not observed as food concentration changed, it is possible that other mechanisms were employed to maximize energy uptake. Specifically, changes in digestive processes may contribute to constant levels of energy absorption despite variability in the quantity and quality of diet in the digestive system [54,73–75]. For example, changing in digestive enzyme activity may increase the absorption efficiency of bivalves acclimated to low-quality diets [76]. In addition, gut passage time may increase in response to diets of low quality to prolong the time available for digestion and absorption of nutrients [76]. The relationships between digestive processes and diet quantity and quality are complex, particularly as natural diets may fluctuate on both short- and long-term timescales; however, they have been empirically modeled [18,77,78]. Changes in digestive processes may contribute to stable energy uptake, despite variations in IR.

5. Conclusions

This study examined the functional relationships between pumping and the ingestion rate in *M. edulis* in response to changes in the diet concentration in a low-seston environment. Results indicated that there were no clear relationships between the population-level pumping rate and food concentration, measured as fluorescence; however, the ingestion rate increased with increasing food concentration. Using novel methodology that permitted the measurement of feeding activity with high temporal resolution, approximately every 20 min, this study highlights the variability in physiological rates both between and within individuals exposed to the same environmental conditions. Both intra- and interindividual variability in pumping and ingestion ranges were observed in all experiments. Understanding the range of both intra- and interindividual variability in physiological rates is beneficial when scaling physiological rates from the individual to population level and for estimating interactions between suspension-feeders and food source. This study contributes to our knowledge of how bivalves acquire energy in dynamic food environments.

Author Contributions: Conceptualization, methodology, validation, all authors.; resources, R.F., T.S., Ø.S.; data curation, L.S., A.A.; writing–original draft preparation, L.S.; writing–reviewing and editing, all authors.; visualization, L.S.; supervision, R.F., Ø.S.; funding acquisition, R.F., Ø.S. All authors have read and agreed to the published version of the manuscript.

Funding: This work was supported by a graduate Natural Sciences and Engineering Research Council of Canada (NSERC) award, a Killam Predoctoral Scholarship, and a Mitacs Globalink Research Award to L.S., an NSERC Discovery Grant to R.F., and the project "Sustainable Low-Trophic Aquaculture" at the Institute of Marine Research (Bergen, Norway).

Institutional Review Board Statement: Not applicable.

Informed Consent Statement: Not applicable.

Data Availability Statement: Not applicable.

Acknowledgments: Authors thank Cathinka Krogness, Justin Trueman, and Tom Koppenol for their assistance with data collection.

Conflicts of Interest: The authors declare no conflict of interest.

References

1. Prins, T.C.; Smaal, A.C.; Dame, R.F. A review of the feedbacks between bivalve grazing and ecosystem processes. *Aquat. Ecol.* **1998**, *31*, 349–359. [CrossRef]
2. Newell, R. Ecosystem influences of natural and cultivated populations of suspension-feeding bivalve molluscs: A review. *J. Shellfish Res.* **2004**, *23*, 51–61.
3. Trottet, A.; Roy, S.; Tamigneaux, E.; Lovejoy, C.; Tremblay, R. Impact of suspended mussels (*Mytilus edulis* L.) on plankton communities in a Magdalen Islands lagoon (Québec, Canada): A mesocosm approach. *J. Exp. Mar. Biol. Ecol.* **2008**, *365*, 103–115. [CrossRef]
4. Smaal, A.C.; Verbagen, J.H.G.; Coosen, J.; Haas, H.A. Interaction between seston quantity and quality and benthic suspension feeders in the Oosterschelde, the Netherlands. *Ophelia* **1986**, *26*, 385–399. [CrossRef]
5. Smaal, A.C.; Schellekens, T.; van Stralen, M.R.; Kromkamp, J.C. Decrease of the carrying capacity of the Oosterschelde estuary (SW Delta, NL) for bivalve filter feeders due to overgrazing? *Aquaculture* **2013**, *404–405*, 28–34. [CrossRef]
6. Bratbak, G.; Jacquet, S.; Larsen, A.; Pettersson, L.H.; Sazhin, A.F.; Thyrhaug, R. The plankton community in Norwegian coastal waters—abundance, composition, spatial distribution and diel variation. *Cont. Shelf Res.* **2011**, *31*, 1500–1514. [CrossRef]
7. Foster-Smith, R.L. The effect of concentration of suspension on the filtration rates and pseudofaecal production for *Mytilus edulis* L., *Cerastoderma edule* (L.) and *Venerupis pullastra* (Montagu). *J. Exp. Mar. Biol. Ecol.* **1975**, *17*, 1–22. [CrossRef]
8. Shumway, S.E.; Cucci, T.L.; Newell, R.C.; Yentsch, C.M. Particle selection, ingestion, and absorption in filter-feeding bivalves. *J. Exp. Mar. Biol. Ecol.* **1985**, *91*, 77–92. [CrossRef]
9. Velasco, L.; Navarro, J. Feeding physiology of infaunal (*Mulinia edulis*) and epifaunal (*Mytilus chilensis*) bivalves under a wide range of concentrations and qualities of seston. *Mar. Ecol. Prog. Ser.* **2002**, *240*, 143–155. [CrossRef]
10. Bayne, B.L.; Iglesias, J.I.P.; Hawkins, A.J.S.; Navarro, E.; Heral, M.; Deslous-Paoli, J.M. Feeding behaviour of the mussel, *Mytilus edulis*: Responses to variations in quantity and organic content of the seston. *J. Mar. Biol. Assoc. UK* **1993**, *73*, 813–829. [CrossRef]
11. Smaal, A.C.; Vonck, A.P.M.A.; Bakker, M. Seasonal Variation in Physiological Energetics of *Mytilus edulis* and *Cerastoderma edule* of Different Size Classes. *J. Mar. Biol. Assoc. UK* **1997**, *77*, 817–838. [CrossRef]

12. Strohmeier, T.; Strand, Ø.; Cranford, P. Clearance rates of the great scallop (*Pecten maximus*) and blue mussel (*Mytilus edulis*) at low natural seston concentrations. *Mar. Biol.* **2009**, *156*, 1781–1795. [CrossRef]

13. Riisgard, H.U. Filtration rate and growth in the blue mussel, *Mytilus edulis* Linneaus, 1758: Dependence on algal concentration. *J. Shellfish Res.* **1991**, *10*, 29–35.

14. Clausen, I.; Riisgård, H. Growth, filtration and respiration in the mussel *Mytilus edulis*: No evidence for physiological regulation of the filter-pump to nutritional needs. *Mar. Ecol. Prog. Ser.* **1996**, *141*, 37–45. [CrossRef]

15. Hawkins, A.; Smith, R.; Bayne, B.; Héral, M. Novel observations underlying the fast growth of suspension-feeding shellfish in turbid environments: *Mytilus edulis*. *Mar. Ecol. Prog. Ser.* **1996**, *131*, 179–190. [CrossRef]

16. Navarro, E.; Iglesias, J.I.P.; Ortega, M.M. Natural sediment as a food source for the cockle *Cerastoderma edule* (L.): Effect of variable particle concentration on feeding, digestion and the scope for growth. *J. Exp. Mar. Biol. Ecol.* **1992**, *156*, 69–87. [CrossRef]

17. Velasco, L.; Navarro, J. Feeding physiology of two bivalves under laboratory and field conditions in response to variable food concentrations. *Mar. Ecol. Prog. Ser.* **2005**, *291*, 115–124. [CrossRef]

18. Willows Optimal digestive investment: A model for filter feeders experiencing variable diets. *Am. Soc. Limnol. Oceanogr.* **1992**, *37*, 829–847. [CrossRef]

19. Rueda, J.L.; Smaal, A.C. Physiological response of Spisula subtruncata (da Costa, 1778) to different seston quantity and quality. In *Nutrients and Eutrophication in Estuaries and Coastal Waters*; Orive, E., Elliott, M., de Jonge, V.N., Eds.; Springer Science & Business Media: Berlin/Heidelberg, Germany, 2002; pp. 505–511.

20. Widdows, J.; Fieth, P.; Worrall, C.M. Relationships between seston, available food and feeding activity in the common mussel *Mytilus edulis*. *Mar. Biol.* **1979**, *50*, 195–207. [CrossRef]

21. Jøsrgensen, C.B. Bivalve filter feeding revisited. *Mar. Ecol. Prog. Ser.* **1996**, *142*, 297–302.

22. Riisgård, H.U.; Egede, P.P.; Barreiro Saavedra, I. Feeding Behaviour of the Mussel, *Mytilus edulis*: New Observations, with a Minireview of Current Knowledge. *J. Mar. Biol.* **2011**, *2011*, 312459. [CrossRef]

23. Winter, J. A critical review on some aspects of filter-feeding in lamellibranchiate bivalves. *Haliotis* **1976**, *7*, 1–87.

24. Steeves, L.; Strohmeier, T.; Filgueira, R.; Strand, Ø. Exploring feeding physiology of *Mytilus edulis* across geographic and fjord gradients in low-seston environments. *Mar. Ecol. Prog. Ser.* **2020**, *651*, 71–84. [CrossRef]

25. Navarro, J.M.; Winter, J.E. Ingestion rate, assimilation efficiency and energy balance in Mytilus chilensis in relation to body size and different algal concentrations. *Mar. Biol.* **1982**, *67*, 255–266. [CrossRef]

26. Navarro, J.; Widdows, J. Feeding physiology of *Cerastoderma edule* in response to a wide range of seston concentrations. *Mar. Ecol. Prog. Ser.* **1997**, *152*, 175–186. [CrossRef]

27. Holling, C.S. The Functional Response of Invertebrate Predators to Prey Density. *Mem. Entomol. Soc. Can.* **1966**, *98*, 5–86. [CrossRef]

28. Picoche, C.; Le Gendre, R.; Flye-Sainte-Marie, J.; Françoise, S.; Maheux, F.; Simon, B.; Gangnery, A. Towards the Determination of *Mytilus edulis* Food Preferences Using the Dynamic Energy Budget (DEB) Theory. *PLoS ONE* **2014**, *9*, e109796. [CrossRef]

29. Montalto, V.; Martinez, M.; Rinaldi, A.; Sarà, G.; Mirto, S. The effect of the quality of diet on the functional response of *Mytilus galloprovincialis* (Lamarck, 1819): Implications for integrated multitrophic aquaculture (IMTA) and marine spatial planning. *Aquaculture* **2017**, *468*, 371–377. [CrossRef]

30. Figueiras, F.; Labarta, U.; Reiriz, M. Coastal upwelling, primary production and mussel growth in the Rías Baixas of Galicia. In *Sustainable Increase of Marine Harvesting: Fundamental Mechanisms and New Concepts*; Springer: Berlin/Heidelberg, Germany, 2002; pp. 121–131.

31. Erga, S.R. Ecological studies on the phytoplankton of Boknafjorden, western Norway. II. Environmental control of photosynthesis. *J. Plankton Res.* **1989**, *11*, 785–812. [CrossRef]

32. Frette, Ø.; Rune Erga, S.; Hamre, B.; Aure, J.; Stamnes, J.J. Seasonal variability in inherent optical properties in a western Norwegian fjord. *Sarsia* **2004**, *89*, 276–291. [CrossRef]

33. Cranford, P.J.; Grant, J. Particle clearance and absorption of phytoplankton and detritus by the sea scallop *Placopecten magellanicus* (Gmelin). *J. Exp. Mar. Biol. Ecol.* **1990**, *137*, 105–121. [CrossRef]

34. Vajedsamiei, J.; Melzner, F.; Raatz, M.; Kiko, R.; Khosravi, M.; Pansch, C. Simultaneous recording of filtration and respiration in marine organisms in response to short-term environmental variability. *Limnol. Oceanogr. Methods* **2021**, *19*, 196–209. [CrossRef]

35. Palmer, R.E.; Williams, L.G. Effect of particle concentration on filtration efficiency of the bay scallop *Argopecten irradians* and the oyster *Crassostrea virginica*. *Ophelia* **1980**, *19*, 163–174. [CrossRef]

36. Filgueira, R.; Labarta, U.; Fernandez-Reiriz, M.J. Flow-through chamber method for clearance rate measurements in bivalves: Design and validation of individual chambers and mesocosm: CR protocol for validation. *Limnol. Oceanogr. Methods* **2006**, *4*, 284–292. [CrossRef]

37. Strohmeier, T.; Strand, ø.; Alunno-Bruscia, M.; Duinker, A.; Rosland, R.; Aure, J.; Erga, S.; Naustvoll, L.; Jansen, H.; Cranford, P. Response of *Mytilus edulis* to enhanced phytoplankton availability by controlled upwelling in an oligotrophic fjord. *Mar. Ecol. Prog. Ser.* **2015**, *518*, 139–152. [CrossRef]

38. Steeves, L.; Vimond, C.; Strohmeier, T.; Casas, S.; Strand, Ø.; Comeau, L.; Filgueira, R. Relationship between pumping rate and particle capture efficiency in three species of bivalves. *Mar. Ecol. Prog. Ser.* **2022**, *691*, 55–68. [CrossRef]

39. Sunde, B.K. *Gill and Labial Palp Areas in Blue Mussels (Mytilus edulis) at Sites with Different Food Quantity*; University of Bergen: Bergen, Norway, 2013.

40. Kooijman, S. *Dynamic Energy Budget Theory for Metabolic Organisation*; Cambridge University Press: Cambridge, UK, 2010; ISBN 0-521-13191-X.

41. Cleveland, W.S.; Devlin, S.J. Locally Weighted Regression: An Approach to Regression Analysis by Local Fitting. *J. Am. Stat. Assoc.* **1988**, *83*, 596–610. [CrossRef]

42. Cranford, P.; Strohmeier, T.; Filgueira, R.; Strand, Ø. Potential methodological influences on the determination of particle retention efficiency by suspension feeders: *Mytilus edulis* and *Ciona intestinalis*. *Aquat. Biol.* **2016**, *25*, 61–73. [CrossRef]

43. Filgueira, R.; Fernandez-Reiriz, M.J.; Labarta, U. Clearance rate of the mussel Mytilus galloprovincialis. I. Response to extreme chlorophyll ranges. *Cienc. Mar.* **2009**, *35*, 405–417. [CrossRef]

44. Pascoe, P.; Parry, H.; Hawkins, A. Observations on the measurement and interpretation of clearance rate variations in suspension-feeding bivalve shellfish. *Aquat. Biol.* **2009**, *6*, 181–190. [CrossRef]

45. Hawkins, A.J.S.; Bayne, B.L.; Bougrier, S.; Heral, M.; Iglesias, J.I.P.; Navarro, E.; Smith, R.F.M.; Urrutia, M.B. Some general relationships in comparing the feeding physiology of suspension-feeding bivalve molluscs. *J. Exp. Mar. Biol. Ecol.* **1998**, *17*, 87–103. [CrossRef]

46. Pouvreau, S.; Bodoy, A.; Buestel, D. In situ suspension feeding behaviour of the pearl oyster, Pinctada margaritifera: Combined effects of body size and weather-related seston composition. *Aquaculture* **2000**, *181*, 91–113. [CrossRef]

47. Pouvreau, S.; Jonquières, G.; Buestel, D. Filtration by the pearl oyster, Pinctada margaritifera, under conditions of low seston load and small particle size in a tropical lagoon habitat. *Aquaculture* **1999**, *176*, 295–314. [CrossRef]

48. Cranford, P.J.; Ward, J.E.; Shumway, S.E. Bivalve Filter Feeding: Variability and Limits of the Aquaculture Biofilter. In *Shellfish Aquaculture and the Environment*; Shumway, S.E., Ed.; Wiley-Blackwell: Oxford, UK, 2011; pp. 81–124. ISBN 978-047-096-096-7.

49. Filgueira, R. Clearance rate of the mussel Mytilus galloprovincialis. II. Response to uncorrelated seston variables (quantity, quality, and chlorophyll content). *Cienc. Mar.* **2010**, *36*, 15–28. [CrossRef]

50. Hawkins, A.; James, M.; Hickman, R.; Hatton, S.; Weatherhead, M. Modelling of suspension-feeding and growth in the green-lipped mussel Perna canaliculus exposed to natural and experimental variations of seston availability in the Marlborough Sounds, New Zealand. *Mar. Ecol. Prog. Ser.* **1999**, *191*, 217–232. [CrossRef]

51. Iglesias, J.I.P.; Navarro, E.; Alvarez Jorna, P.; Armentina, I. Feeding, particle selection and absorption in cockles Cerastoderma edule (L.) exposed to variable conditions of food concentration and quality. *J. Exp. Mar. Biol. Ecol.* **1992**, *162*, 177–198. [CrossRef]

52. Bayne, B.; Hawkins, A.; Navarro, E.; Iglesias, I. Effects of seston concentration on feeding, digestion and growth in the mussel Mytilus *edulis*. *Mar. Ecol. Prog. Ser.* **1989**, *55*, 47–54. [CrossRef]

53. Hawkins, A.J.S.; Bayne, B.L. Seasonal variation in the balance between physiological mechanisms of feeding and digestion in *Mytilus edulis* (Bivalvia: Mollusca). *Mar. Biol.* **1984**, *82*, 233–240. [CrossRef]

54. Bayne, B.L.; Hawkins, A.J.S.; Navarro, E. Feeding and Digestion by the Mussel *Mytilus edulis* L. (Bivalvia: Mollusca) in Mixtures of Silt and Algal Cells at Low Concentrations. *J. Exp. Mar. Biol. Ecol.* **1987**, *111*, e0205981. [CrossRef]

55. Cranford, P.J.; Emerson, C.W.; Hargrave, B.T.; Milligan, T.G. In Situ Feeding and Absorption Responses of Sea Scallops Placopecten Magellanicus (Gmelin) to Storm-Induced Changes in the Quantity and Composition of the Seston. *J. Exp. Mar. Biol. Ecol.* **1998**, *219*, 45–70. [CrossRef]

56. Frechette, M.; Bourget, E. Significance of Small-Scale Spatio-Temporal Heterogeneity in Phytoplankton Abundance for Energy Flow in Mytilus Edulis. *Mar. Biol.* **1987**, *94*, 231–240. [CrossRef]

57. Bayne, B.L.; Svensson, S.; Nell, J.A. The Physiological Basis for Faster Growth in the Sydney Rock Oyster, Saccostrea Commercialis. *Biol. Bull.* **1999**, *197*, 377–387. [CrossRef] [PubMed]

58. Fuentes-Santos, I.; Labarta, U.; Fernández-Reiriz, M.J. Characterizing Individual Variability in Mussel (*Mytilus galloprovincialis*) Growth and Testing Its Physiological Drivers Using Functional Data Analysis. *PLoS ONE* **2018**, *13*, e0205981. [CrossRef] [PubMed]

59. Tamayo, D.; Ibarrola, I.; Urrutia, M.B.; Navarro, E. The Physiological Basis for Inter-Individual Growth Variability in the Spat of Clams (*Ruditapes philippinarum*). *Aquaculture* **2011**, *321*, 113–120. [CrossRef]

60. Hawkins, A.J.S.; Magoulas, A.; Héral, M.; Bougrier, S.; Naciri-Graven, Y.; Day, A.J.; Kotoulas, G. Separate Effects of Triploidy, Parentage and Genomic Diversity upon Feeding Behaviour, Metabolic Efficiency and Net Energy Balance in the Pacific Oyster *Crassostrea gigas*. *Genet. Res.* **2000**, *76*, 273–284. [CrossRef]

61. Fernández-Reiriz, M.; Garrido, J.; Irisarri, J. Fatty Acid Composition in Mytilus Galloprovincialis Organs: Trophic Interactions, Sexual Differences and Differential Anatomical Distribution. *Mar. Ecol. Prog. Ser.* **2015**, *528*, 221–234. [CrossRef]

62. Griffith, A.W.; Gobler, C.J. Transgenerational Exposure of North Atlantic Bivalves to Ocean Acidification Renders Offspring More Vulnerable to Low PH and Additional Stressors. *Sci. Rep.* **2017**, *7*, 11394. [CrossRef]

63. Zhang, F.; Hu, B.; Fu, H.; Jiao, Z.; Li, Q.; Liu, S. Comparative Transcriptome Analysis Reveals Molecular Basis Underlying Fast Growth of the Selectively Bred Pacific Oyster, Crassostrea Gigas. *Front. Genet.* **2019**, *10*, 610. [CrossRef]

64. de Villemereuil, P.; Gaggiotti, O.; Mouterde, M.; Till-Bottraud, I. Common Garden Experiments in the Genomic Era: New Perspectives and Opportunities. *Heredity* **2016**, *116*, 249–254. [CrossRef]

65. Jørgensen, C.; Larsen, P.; Riisgård, H. Effects of Temperature on the Mussel Pump. *Mar. Ecol. Prog. Ser.* **1990**, *64*, 89–97. [CrossRef]

66. Rosa, M.; Ward, J.E.; Shumway, S.E. Selective Capture and Ingestion of Particles by Suspension-Feeding Bivalve Molluscs: A Review. *J. Shellfish Res.* **2018**, *37*, 727–746. [CrossRef]

67. Coughlan, J. The Estimation of Filtering Rate from the Clearance of Suspensions. *Mar. Biol.* **1969**, *2*, 356–358. [CrossRef]

68. Møhlenberg, F.; Riisgård, H.U. Efficiency of Particle Retention in 13 Species of Suspension Feeding Bivalves. *Ophelia* **1978**, *17*, 239–246. [CrossRef]
69. Rosa, M.; Ward, J.E.; Holohan, B.A.; Shumway, S.E.; Wikfors, G.H. Physicochemical Surface Properties of Microalgae and Their Combined Effects on Particle Selection by Suspension-Feeding Bivalve Molluscs. *J. Exp. Mar. Biol. Ecol.* **2017**, *486*, 59–68. [CrossRef]
70. Pales Espinosa, E.; Perrigault, M.; Ward, J.E.; Shumway, S.E.; Allam, B. Lectins Associated With the Feeding Organs of the Oyster *Crassostrea Virginica* Can Mediate Particle Selection. *Biol. Bull.* **2009**, *217*, 130–141. [CrossRef] [PubMed]
71. Yahel, G.; Marie, D.; Beninger, P.; Eckstein, S.; Genin, A. In Situ Evidence for Pre-Capture Qualitative Selection in the Tropical Bivalve Lithophaga Simplex. *Aquat. Biol.* **2009**, *6*, 235–246. [CrossRef]
72. Strohmeier, T.; Strand, Ø.; Alunno-Bruscia, M.; Duinker, A.; Cranford, P.J. Variability in Particle Retention Efficiency by the Mussel Mytilus Edulis. *J. Exp. Mar. Biol. Ecol.* **2012**, *412*, 96–102. [CrossRef]
73. Bayne, B.L.; Hawkins, A.J.S.; Navarro, E. Feeding and Digestion in Suspension-Feeding Bivalve Molluscs: The Relevance of Physiological Compensations. *Am Zool* **1988**, *28*, 147–159. [CrossRef]
74. Ibarrola, I.; Larretxea, X.; Iglesias, J.I.P.; Urrutia, M.B.; Navarro, E. Seasonal Variation of Digestive Enzyme Activities in the Digestive Gland and the Crystalline Style of the Common Cockle Cerastoderma Edule. *Comp. Biochem. Physiol.* **1998**, 25–34. [CrossRef]
75. Navarro, E.; Iglesias, J.I.P.; Ortega, M.M.; Larretxea, X. The Basis for a Functional Response to Variable Food Quantity and Quality in Cockles Cerastoderma Edule (Bivalvia, Cardiidae). *Physiol. Zool.* **1994**, *67*, 468–496. [CrossRef]
76. Ibarrola, I.; Navarro, E.; Iglesias, J.I.P. Short-Term Adaptation of Digestive Processes in the Cockle Cerastoderma Edule Exposed to Different Food Quantity and Quality. *J. Comp. Physiol. B Biochem. Syst. Environ. Physiol.* **1998**, *168*, 32–40. [CrossRef]
77. Scholten, H.; Smaal, A.C. Responses of Mytilus Edulis L. to Varying Food Concentrations: Testing EMMY, an Ecophysiological Model. *J. Exp. Mar. Biol. Ecol.* **1998**, *219*, 217–239. [CrossRef]
78. Scholten, H.; Smaal, A.C. The Ecophysiological Response of Mussels (Mytilus Edulis) in Mesocosms to a Range of Inorganic Nutrient Loads: Simulations with the Model EMMY. *Aquat. Ecol.* **1999**, *33*, 83–100. [CrossRef]

Journal of
Marine Science and Engineering

MDPI

Article

Filtration Rates and Scaling in Demosponges

Hans Ulrik Riisgård [1],* and Poul S. Larsen [2]

[1] Marine Biological Research Centre, University of Southern Denmark, 5300 Kerteminde, Denmark
[2] Department of Mechanical Engineering, Technical University of Denmark, 2800 Lyngby, Denmark; psl@mek.dtu.dk
* Correspondence: hur@biology.sdu.dk

Abstract: Demosponges are modular filter-feeding organisms that are made up of aquiferous units or modules with one osculum per module. Such modules may grow to reach a maximal size. Various demosponge species show a high degree of morphological complexity, which makes it difficult to classify and scale them regarding filtration rate versus sponge size. In this regard, we distinguish between: (i) small single-osculum sponges consisting of one aquiferous module, which includes very small explants and larger explants; (ii) multi-oscula sponges consisting of many modules, each with a separate osculum leading to the ambient; and (iii) large single-osculum sponges composed of many aquiferous modules, each with an exhalant opening (true osculum) leading into a common large spongocoel (atrium), which opens to the ambient via a static pseudo-osculum. We found the theoretical scaling relation between the filtration rate (F) versus volume (V) for (i) a single-osculum demosponge to be $F = a_3 V^{2/3}$, and hence the volume-specific filtration rate to scale as $F/V \approx V^{-1/3}$. This relation is partly supported by experimental data for explants of *Halichondria panicea*, showing $F/V = 2.66 V^{-0.41}$. However, for multi-oscula sponges, many of their modules may have reached their maximal size and hence their maximal filtration rate, which would imply the scaling $F/V \approx$ constant. A similar scaling would be expected for large pseudo-osculum sponges, provided their volume was taken to be the structural tissue volume that holds the pumping units, and not the total volume that includes the large atrium volume of water. This may explain the hitherto confusing picture that has emerged from the power-law correlation ($F/V = aV^{b}$) of many various types of demosponges that show a range of negative b-exponents. The observed sharp decline in the volume-specific filtration rate of demosponges from their very small to larger sizes is discussed.

Keywords: allometric scaling; sponge module; choanocyte density; specific filtration rate

Citation: Riisgård, H.U.; Larsen, P.S. Filtration Rates and Scaling in Demosponges. *J. Mar. Sci. Eng.* **2022**, *10*, 643. https://doi.org/10.3390/jmse10050643

Academic Editor: Azizur Rahman

Received: 29 March 2022
Accepted: 6 May 2022
Published: 8 May 2022

Publisher's Note: MDPI stays neutral with regard to jurisdictional claims in published maps and institutional affiliations.

1. Introduction

There are nearly 9500 living species of sponges, and the class of demosponges contains 82% of all sponge species [1]. All demosponges are modular filter-feeding organisms that are made up of aquiferous units or modules with one osculum per module [2,3]. The many different species of demosponges show a high degree of morphological complexity [4]. Therefore, they are not easy to classify and scale regarding basic features, such as filtration rate versus sponge size. In the present study, we distinguish between: (i) small single-osculum single-module sponges consisting of one aquiferous module, which includes very small [5] and larger explants [6] (Figure 1); (ii) multi-oscula multi-modular sponges consisting of many aquiferous modules each with a separate osculum leading to the ambient, which could be small explants [7] or larger sponges, such as *Halichondria panicea* [8–10]; and (iii) large single-osculum multi-modular sponges (or large single-pseudo-osculum sponges) composed of many aquiferous modules each with an exhalant opening (true osculum) leading to a common large spongocoel (atrium), which opens to the ambient via a static pseudo-osculum, such as *Xestospongia muta* [11,12]. A contraction of the true oscula in the atrial lining of *Verongia gigantea* was described by [13] and only "very small specimens" with a body volume <200 mL were able to occlude the joint pseudo-osculum.

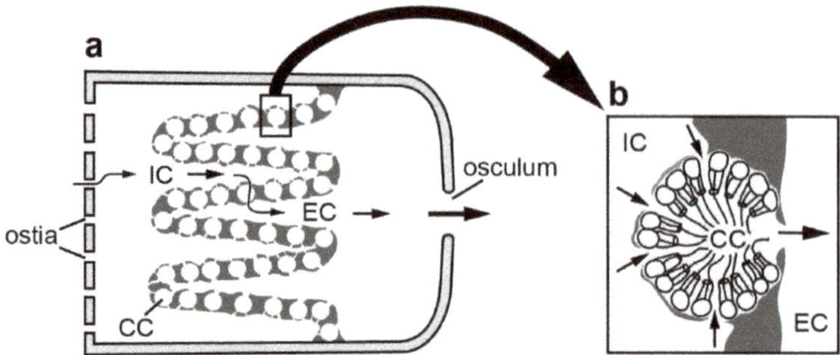

Figure 1. Sketch of a single-osculum demosponge showing the water flow from outside through ostia into inhalant canals (ICs) to the water pumping choanocyte chambers (CCs), where the water is filtered, then further into exhalant canals (ECs) and subsequently out through the osculum. The thin wall separating the tapered inhalant and exhalant canals is for a great part made up of CCs embedded in the mesenchyme. Adapted from [9].

Single-osculum and multi-oscula sponges have contraction–inflation behavior, including the closure and opening of the osculum; furthermore, in these sponges, the speed of the exhalant jet correlates to the size of the osculum [5,6,14]. Here, [6] suggested the following theoretical allometric scaling parameters for the osculum jet speed (U), osculum cross-sectional area (OSA), and pumping rate (=filtration rate, F) could be expressed as:

$$U = a_1 OSA^{b1}; \; b_1 = 1/2 \tag{1}$$

$$F = a_2 OSA^{b2}; \; b_2 = 3/2 \tag{2}$$

These scaling parameters, which rely on the suggested uniform density of pumping units (choanocyte chambers), were found to agree with the measurements on both single-osculum explants [6] and multi-oscula explants of *Halichondria panicea* [7]. However, to examine how the theoretical scaling relationships applies to larger sponges, [15] measured in situ the filtration rate of 20 sponge species and found that their results showed "an opposite trend of an allometric decrease in U with OSA for two-thirds (12 out of 18) of the species", and they concluded that the allometric scaling parameters did not apply to large sponges with "fully open and static oscula", but only to small explants that "dynamically constrict and expand their oscula" [15] found that their data showed a different scaling than that of Equations (1) and (2), because the decrease in volume-specific pumping rate with increasing sponge size for the larger sponges indicated that the density of choanocyte chambers decreases with increasing sponge volume.

To help in the understanding of the allometric data correlations [16], the compiled available data on the volume-specific filtration rate (F/V) versus sponge volume (V) approximated as $F/V \approx V^b$ in demosponges, but the observed large and confusing variations could not be immediately explained. Therefore, an important aspect of the present study is to clear up this situation. It is our hypothesis that F/V versus V of (i) single-osculum single-module demosponges decreases with increasing size, while it remains essentially constant for (ii) multi- and (iii) single-oscular multi-modular demosponges, provided that volume is considered to be that of structural sponge tissue, which, again, is proportional to the sponge dry weight.

Here, we first examine the scaling relation between the filtration rate and body volume in (i) single-osculum single-module explants before we compare it with scaling in (ii) multi-oscular multi-modular sponges. Next, we examine if (iii) a large single-osculum multi-modular sponge may be regarded as a population of modules that share the characteristics of single-osculum explants, or if such sponges have different scaling characteristics. Finally, we discuss how to arrive at a better understanding of specific filtration rates in demosponges. We arrive at the classification of demosponges from the concept of "modules", which is then used in the scaling of the filtration rates.

2. Materials and Methods

We used published data on single-osculum sponge explants consisting of one aquiferous module of various sizes obtained from colonies of the demosponge *Halichondria panicea*. Branches of the collected sponges were either cut into very small pieces without an osculum [5] or in fragments of various sizes with a single osculum [6]. The cut-off pieces were individually fixed with whipping twine on substrate plates in flowing seawater and were allowed to develop into explants over a couple of weeks, which reorganized their elements of the aquiferous system [3] in such a way that each osculum cross-sectional area (*OSA*) became adjusted to the size (volume, *V*) of the individual sponge explant. The experimental data obtained for these explants at 15 °C were used to scale the filtration rate with the size of the sponge module. Due to the very low volume-specific filtration rates in the single-osculum explants reported by [7], we suggest that these explants may not have been fully reorganized, and therefore not used in the present study. Power-function regression curves (LM) were fitted the in [17] for growth rate estimates, based on the sponge body volume over time.

3. Results and Discussion

In the present study, the scaling relation between the filtration rate and volume of single-osculum explants is presented and compared with scaling in multi-oscular sponges. The findings are discussed in order to obtain a better understanding of how to deal with specific filtration rates in demosponges.

3.1. Scaling in Single-Osculum Single-Module Demosponges

A scaling relation between the water-pumping rate and sponge-body volume may be derived by considering a single inhalant canal of length L in a single-osculum demosponge (Figure 1). The thin wall separating the tapered inhalant and exhalant canal system is for a great part (30% to 50%) made up of water-pumping choanocyte chambers with a diameter of approximately 30 μm embedded in mesenchyme. The pumping rate (=filtration rate, F) from these chambers is proportional to the product of pumping rate (F_{CC}) of each choanocyte chamber and their number, which is proportional to the wall area ($\sim L^2$) of the canal, i.e., $F \approx L^2$, and the sponge volume associated with canals and walls would scale as $V \sim L^3$ for the isometric growth. It follows that $F \approx (V^{1/3})^2$ and thus:

$$F = a_3 V^{b3}; b_3 = 2/3 \tag{3}$$

Hence, the volume-specific filtration rate would scale as $F/V = V^{2/3-1} = V^{-1/3}$, which indicates a decrease with increasing sponge volume. This scaling may be expected to apply when a small single-osculum sponge grows bigger. Thus, [5] measured the filtration rate in 15 small single-osculum *Halichondria panicea* explants of the same size ($V = 0.018$ mL) and found that the mean filtration rate was $F = 0.28 \pm 0.06$ mL min^{-1}, which indicates a volume-specific filtration rate of $F/V = 0.28/0.018 = 15.6$ min^{-1}, thus showing that the explant filters an amount of water that is equivalent to 15.6 times its body volume per min. Using Equation (3) $F = a_3 V^{2/3}$ the filtration rate (F, mL min^{-1}) versus sponge body volume (V, mL) can be predicted to be $F = 3.97 V^{2/3}$ because $a_3 = F/V^{2/3} = 0.28/0.018^{2/3} = 3.97$ and consequently caused the volume-specific filtration rate to be $F/V = 3.97 V^{2/3-1} = 3.97 V^{-1/3}$. The predicted F/V versus V is depicted in Figure 2. Furthermore, [5] measured F/V in a number of single-osculum *H. panicea* explants with various body sizes ($V = 0.018$ to 1.977 mL), which are also shown in Figure 2. It can be observed that the model-predicted curve describes the experimental data fairly well. Another example of a single aquiferous module is the sponge branch cut from a colony of *Haliclona urceolus* [9], for which $F = 6$ mL min^{-1} and $V = 1.726$ mL was measured. These results lead to $F/V = 3.48$ min^{-1} and in are good agreement with the foregoing example, which implies $a_3 = F/V^{2/3} = 4.17$.

Furthermore, the regression analysis of the measured filtration rate and size of 8 *H. urceolus* specimens [8] produced $F/V = 3.96V^{-0.39}$. Likewise, the exponent ($b_3 = 0.59$) for the power function regression line for F versus V for the same data is close to the model-predicted (Equation (1)) $b_3 = 0.66$ (Figure 3). We should add that the same results from [6] were shown in [7], where, we as co-authors, erroneously assume the linear scaling relation $F = aV$ now replaced by $F = 2.66 V^{0.59}$. The same mistake was made in [7].

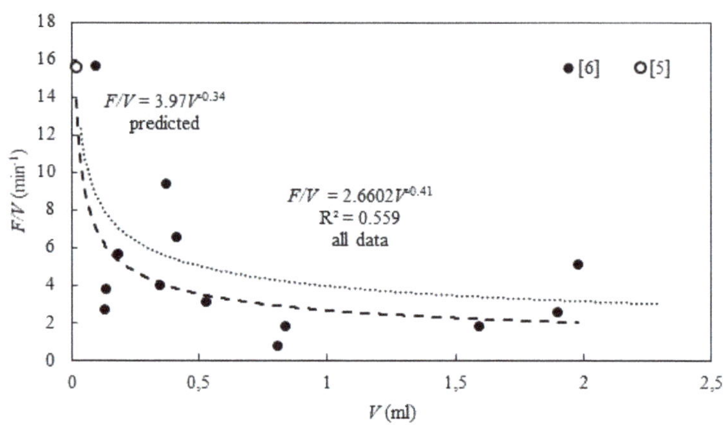

Figure 2. *Halichondria panicea.* Volume-specific filtration rate (F/V) as a function of body volume (V) of single-osculum explants. The model-predicted curve (dotted) based on [5] (open symbol) is shown along with the power-function regression line for all data (dashed, solid symbols) for data obtained from [6] (LM, $t_{0.2969, 12} = -2.859$, $p = 0.0013$).

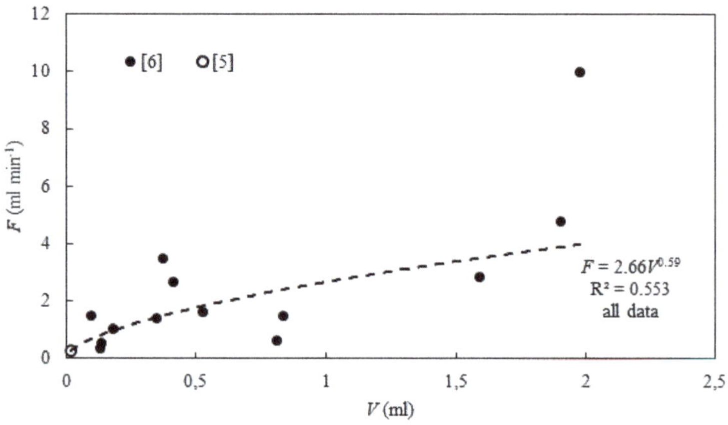

Figure 3. *Halichondria panicea.* Filtration rate (F) of single-osculum explants as a function of sponge-body volume (V). The power-function regression line has been shown along with its equation. The b_3-exponent is 0.59, which may be compared to the model-predicted $b_3 = 2/3$ [5,6] (LM, $t_{0.2964, 12} = 4.099$, $p = 0.0015$).

An aquiferous module is "a certain volume in the sponge that is supplied by a system of choanocyte chambers and aquiferous canals associated with a single osculum. Therefore, a sponge represents a modular organism" [18]. A demosponge, such as *Halichondria panicea* consists of multiple modules, each with an osculum (Figure 4). If a module is only able to grow until it has obtained a certain volume, most of the whole modular sponge organism will consist of full-grown modules with a near similar *F/V* ratio. Therefore, the *F/V* ratio of a growing multi-oscula sponge in which most of the modules are full-grown should be expected to also be constant. Thus, the present scaling, Equation (3) only applies when a small single-osculum sponge—or an aquiferous module—grows larger.

Figure 4. *Halichondria panicea.* Underwater photo (29 July 2019) from the inlet to Kerteminde Fjord, Denmark, showing erect branching (ramose) sponges of type (ii) with multi-modules each with an osculum.

3.2. Scaling in Multi-Oscula Multi-Modular Demosponges

To our knowledge, the first attempt to outline the size and number of aquiferous modules in a multi-oscular sponge was made by [7] in explants of branching *Halichondria panicea*. Here, the boarders of each module were identified through observations of incurrent and excurrent water flow using fluorescein deposited on the sponge surface (exopinacoderm) using a micromanipulator, and subsequently the volume (V_{mod}) of each module was measured by cutting along the borders and weighing the module. Because the modules were not growing, a plot of *F/V* versus *V* showed no trend, i.e., *F/V* = constant, and for modules of sizes between 0.5 and 2.7 mL, the mean *F/V* was found to be 1.2 ± 0.8 min^{-1} [7]. Similar studies in other multi-oscula multi-modular demosponges are awaiting in order to verify if *F/V* ~ constant.

3.3. Scaling in Single-Osculum Multi-Modular Demosponges

Many large tropical single-osculum sponges are composed of many modules, each with a true osculum that opens into a common spongocoel (atrium), which opens to the ambient water via a fully open motionless pseudo-sculum. Here, *Aplysina lacunosa* may serve as an example of a typically large tropical species, which has a tubular shape with a large pseudo-osculum on top [19]. Another example is *Verongia gigantea* in which the (pseudo)osculum becomes unable to occlude when the diameter becomes greater than about 15 mm, whereas the atrial wall with true oscula shows periodic cessation [13].

For clarity, when dealing with multi-oscular sponges, we define "structural volume V_{str}" as that of the sponge-tissue structure that excludes any large spongocoel (atrium cavity), while the "total volume V_{tot}" of a sponge includes the atrium. Here, V_{str} may be expected to be proportional to the dry weight of sponge tissue, W. Small single-osculum explants have no large atrium, just exhalant canals that join to one canal leading to the osculum; therefore, here, there is no significant difference between the 2 volumes. Likewise, in a multi-oscular sponge, such as *Halichondria panicea*, each module has its own osculum. However, for large vase, jar, urn- or tube-shaped sponges, the 2 volumes may be quite different, as it appears from the following.

In [11], it was found that the relationship between the spongocoel volume (V_{spongo}) and total sponge volume (V_{tot}) could be described by the following allometric equation: $V_{spongo} = \alpha V_{tot}^{1.214}$ where the exponent $\beta = 1.214$ indicates that the relative volume of the spongocoel may increase by as much as 20 to 25% for sponge sizes of 50 to 200 L for the total volume of *Xestospongia muta*. Other β-exponents may apply for other large sponge species with a spongocoel of various shapes, and therefore a scaling of F/V in one sponge species may not apply to another species. Furthermore, [15,20] found in some twenty sponge species that the F/V ratio decreases with increasing V_{tot} calculated from photos, including the spongocoel. Here, it is noteworthy that [21] excluded the spongocoel ("atrial cavity") when the field sizes of the three demosponges *Mycale* sp., *Verongia gigantea*, and *Tethya crypta*, were determined, in which the tissue-volume specific filtration rate was found to be constant, independent of the sponge body size [22]. This trend of constancy is the same as the above suggestion that the F/V ratio of a multi-oscular sponge tends to be constant because all the modules that build up the sponge body are not growing and/or of comparable size with a comparable F/V ratio.

Determining the "flux per unit sponge tissue" for 14 species, [23] found volume-specific filtration rates that were also essentially constant, $b = 0.045$. Data by [24] shows that 24 to 27 °C leads to $b = -0.071$ for *Cinachyrella* cf. *cavernose*, while including their data at 30 to 33 °C leads to $b = 0.23$, which suggest an increase in their size rather than a decrease, which is difficult to explain.

For five Mediterranean sponge species, [20] determined exponents b_4 and b_5 in the correlations $W \sim V_{tot}^{b4}$ and $F \sim V_{tot}^{b5}$ from which we calculated the exponent $b_6 = b_5/b_4 - 1$ in $F/W \sim W^{b6}$ as the values of $b_6 = -0.63, -0.10, -0.17, -0.13,$ and -0.42. The near-zero value of b_6 for three of the sponges (*Crambe crambe*, *Petrosia ficiformis*, and *Chondrosia reniformis*) suggest that they have a nearly constant weight-specific filtration rate. This example shows that an increasing fraction of the sponge volume in these types of sponges is made up of canals with water. Therefore, for these types of sponges, specific filtration rates should be based on the dry weight (W), in which case, the specific filtration rates appear to be nearly independent of size in this sense.

Although many big vase-, jar-, urn-, or barrel-type sponges have only one common exhalant opening (pseudoosculum), these sponges cannot be directly compared to single-osculum modules because a large (unknown) number of modules enter into, e.g., a giant barrel sponge and because, possibly, the majority of these modules have grown to their maximum size. In the study of allometry and scaling of sponges, it appears to be essential to distinguish between types of sponges as described in Sections 1–3, and specifically employ the structural volume in F/V. However, F/W versus W (or biomass, AFDW) for the same species would probably show that F/W = constant.

As can be observed from the literature, there is a considerable interest in estimating the grazing impact from observed populations of given species of sponges at a given site. This has led to the much data on the filtration rate versus size in terms of the volume of various species of demosponges summarized by [16], who compiled available data on volume-specific filtration rate (F/V) versus sponge volume (V) approximated as $F/V = aV^b$ in demosponges. However, the observed variations were large and confusing and could not be immediately explained. However, the present assessment should help to clarify the situation. The scaling represented by Equation (3) does not apply for multi-oscula multi-modular sponges and single-osculum multi-modular demosponges, which explains the confusing picture of the species shown in Figure 1 of [16], in which all the various types of demosponge species have been shown together and approximated by the power-law $F/V = aV^{b1-1} = aV^b$, where b ~ 0 when b_1 is close to 1, but without a more precise definition of V (i.e., sponge-body volume with or without spongocoel).

4. Filtration Rate, *OSA*, and Size

From the foregoing, it appears that a multi-oscula demosponge may be regarded as a population of modules each sharing the characteristics of a single-osculum explant, but also that the scaling of F/V versus V in single-osculum modules does not apply to multi-oscula sponges, which, due to their population of modules, obtained different and scaling characteristics for F/V versus the total sponge volume V.

The scaling relations between the filtration rate and osculum cross-sectional area is of interest because "the number of oscula and their *OSA* were the best predictors" of the filtration rate of sponges [15]. However, again, we must distinguish between the different types of demosponges. Thus, $F = a_2OSA^{b2}$ and $b_2 = 3/2$ in Equation (2) for single-osculum modules but have the values of $b_2 = 0.75$ to 1.07 for single-osculum multi-modular demosponges [15].

As an example of scaling, Figure 4 shows an underwater photo of *Halichondria panicea*, which consists of multi-oscular multi modules, each with an osculum with a mean diameter of 1.9 ± 0.6 mm (29 July 2019), giving rise to $OSA = 2.84$ mm^2, $V = 3.2$ mL, $F = 7.4$ mL min^{-1}, and $F/V = 2.3$ min^{-1} when using Equation (2) $F = a_2OSA^{3/2}$ with $a_2 = 1.55$ [6], and $a_3 = 3.97$ in $F = a_3V^{2/3}$ (Equation (3)) shown in Figure 2. Obviously, the in situ measurement of F and the calibration of the scaling relations to the actual temperature are desirable in order to verify the predictions, but the example illustrates how scaling relations may be useful in field studies because only the dimensions of the oscula need to be measured to obtain information about both the size and filtration rate of each module of a multi-oscula sponge, which can be considered as a population of modules.

The size of an aquiferous sponge module and its *OSA* are closely interconnected and follow a fixed scaling, but the module size and the *OSA* are not stable as evident from the following, where *Halichondria panicea* again serves as an example. An earlier measurement of the average size of *OSA* in *H. panicea* at the same field location (Figure 4) was measured by [7] to the lower value of 1.0 ± 0.6 mm^2 (17 December 2018). Thus, it is likely that the mean size of otherwise full-grown modules and concurrently the *OSA* and F change over the season, along with pronounced changes in the condition index [25,26]. Furthermore, [27] observed that the range of the oscular diameter in *H. panicea* was 1 to 4 mm of sponges from sites with "medium to high current velocities", but smaller, 0.5 to 2 mm, at "low current" sites. From these observations, it can be concluded that the size of modules and their *OSA* vary between localities and over the season, and that the size and shape of the individual modules of *H. panicea* are not stable, but depend on the living site and time of the year. The degree of polymorphism in this sponge is "much higher than observed in most other sponges", and its growth form may be encrusting, lumpy, or ramose (Figure 4), depending on the current velocities and degree of exposure to waves [27].

5. Density and Filtration Rates of Choanocyte Cambers in a Single-Osculum Module

The prerequisite for the scaling leading to Equation (3) is that the "area specific filtration rate" of the thin wall separating the inhalant and exhalant canals is constant, i.e., the wall-specific density of choanocytes (CCs) and their individual filtration rate is constant. The suggested scaling is verified by the experimental data shown in Figure 2. Because of this scaling, the CC density decreases with increasing V, whereas the filtration rate of the single choanocyte may be reduced due to increasing the system resistance when the canals become longer.

Here, it should be mentioned that Figure 1 is very schematic. Thus, although "the exhalant system, in a crude sense, mirrors the inhalant system" [4], the two systems serve different functions. Food particles >5 μm are filtered out of the inflowing water in the inhalant canal system and phagocytosed here, whereas smaller particles are retained in the CCs [28]. The exhalant canal system acts as a sewage system, which carries filtered water, excretion products, and indigestible matter out of the sponge, and it is noteworthy that the diameter of the exhalant apertures are more than two times larger than the inhalant apertures [4,29]. The significance of this difference in the aperture diameter remains unknown, but the resistance to flow may be relatively lower in the larger exhalant canals.

5.1. Filtration Rates

It is our hypothesis that the volume specific filtration rate (F/V) versus sponge volume (V) of single-osculum single-module demosponges decreases with the increasing size, while it remains essentially constant for multi-oscular multi-modular and single-osculum multi-modular demosponges, provided the sponge volume is that of the structural sponge tissue, which, again, is proportional to the sponge dry weight. Furthermore, we seek to understand the cause of the observed strong decrease in F/V with increasing V. The specific filtration rate equals the product of density (n_{CC}) and filtration rate (F_{CC}) of the choanocyte chambers, $F/V = n_{CC} \times F_{CC}$, where each factor may decrease in the process of growth, n_{CC} due to an increasing volume fraction of the tissue, F_{CC} due to the changing choanocyte pump performance related to seal imperfections in pumps and/or increasing pressure losses in the aquiferous system with increasing size.

For 5 sponges, [20] shows very large volume-specific pumping rates (10 to 40 min^{-1}) for small individuals that then decrease with increasing size to more typical values (2 to 6 min^{-1}); however, these results may be subject to corrections for the use of total volume rather than structural tissue volume. Furthermore, based on the decreasing F/V with increasing V observed "in most of the sponges" studied by [15], the authors suggested that the CC density was concurrently reduced. This possibility is now discussed by considering some examples of the demosponge *Halichondria panicea*.

A very small explant of volume $V = 0.018$ mL was found to have the high-volume specific filtration rate of $F/V = 15.6\ \text{min}^{-1}$ [5], while the larger, near full grown explant of volume $V = 1$ mL had the smaller value $F/V = 2.66\ \text{min}^{-1}$ [6]. For an estimate of the order of magnitude of n_{CC} and F_{CC}, we considered a *Halichondria panicea* specimen having $F/V = 6.1\ \text{min}^{-1}$ [30] and $n_{CC} = 18{,}000\ \text{mm}^{-3}$ [4], which produced $F_{CC} = (6.1/18{,}000)/60 = 5.65 \times 10^{-6}\ \text{mm}^3\ \text{s}^{-1} = 5650\ \text{μm}^3\ \text{s}^{-1}$.

As a first scenario, we assumed $F_{CC} = 5650\ \text{μm}^3\ \text{s}^{-1}$ to prevail for the small and the larger explants. This would imply that the choanocyte chamber density should decrease from $n_{CC} = (15.6 \times 10^9/5650)/60 = 46{,}018\ \text{mm}^{-3}$ to $n_{CC} = 7847\ \text{mm}^{-3}$, i.e., by a 5.9 factor, as determined by the ratio of F/V values. Furthermore, assuming a typical chamber diameter of 30 μm, the volume fraction of the chambers would be $\pi/6 \times 30^3 \times 46{,}018 \times 10^{-9} \times 100 = 65\%$ and 11%, respectively, leaving little space for aquiferous canals and other structure in the first case, which can be justified by the very short canals in a very small explant.

The density ($n_{CC} = 18{,}000\ \text{mm}^{-3}$) reported by [4] represents "mature regions with relatively stable dimensions" in *Halichondria panicea*, and the attainment of samples from "growth points" was deliberately avoided. Because the very small explant represents a "growth point", this may suggest a higher chamber density here, but verification awaits further information on the structure and dimensions of the canal system at these "growth points".

With 80 choanocytes in a CC [4], the filtration rate of a single choanocyte in *Halichondria panicea* was estimated at $F_{ch} = (5650/80 =) 70.6 \ \mu m^3 \ s^{-1}$. With 95 choanocytes per CC in *Haliclona permollis* [4] and $F/V = 6.0 \ min^{-1}$ measured in the closely related *H. urceolus* [8], it was estimated that $F_{ch} = (6.0 \times 60)/(12{,}000 \times 95 \times 10^3) = 0.32 \times 10^{-6} \ mL \ h^{-1}$ or 89 $\mu m^3 \ s^{-1}$. For the comparison with other demosponges, [22] found that the tropical demosponge *Tethya crypta* had a volume-specific filtration rate of $F/V = 10.8 \ min^{-1}$, and from this, it was estimated (using CC density and number of choanocytes per CC reported by [31] that $F_{ch} = 648/(14{,}403 \times 99 \times 10^3) = 0.46 \times 10^{-6} \ mL \ h^{-1}$ or 128 $\mu m^3 \ s^{-1}$. Other, but strongly varying, values were calculated by [32]) using the data reported by [31]. Thus, F_{ch} was calculated from the measured F/V ratio divided by the CC density. However, the CC density was determined for sponge tissue, whereas V was "calculated by measuring the dimensions of the sponge from images taken of whole animals in situ" [31], and this may explain some of the strong variations in F_{ch} between species, but also the differences between CC density in "growth points" and "full grown" modules may have contributed to the variation.

In the second scenario, we assume that the CC density was constant, which would lead to a change in the CC filtration rate produced by the factor 5.9 of the F/V ratio of the very small explant to the larger one. Using the aforementioned value of $F_{CC} = 5650 \ \mu m^3 \ s^{-1}$, or $F_{ch} = 70.6 \ \mu m^3 \ s^{-1}$ for 80 choanocytes per CC, for the larger explant, it would suggest a high value of $F_{ch} = 70.6 \times 5.9 = 416 \ \mu m^3 \ s^{-1}$ for the very small explant. Although an increase in F_{ch} would be expected for the much shorter canals in the very small explant, it would be less than suggested here unless the choanocyte pumps at this stage were more efficient. A value of 453 $\mu m^3 \ s^{-1}$ for an opposing system pressure of 1 mm H_2O was computed by [32], provided good seals represented by the second reticulum (acting as a "gasket") and the glycocalyx mesh on the upper part of the collar.

By comparing the two scenarios, it appears that the main contribution to the decrease in the F/V ration during growth of the single-osculum explant module was due to the decrease in the chamber density because of the increase in the volume of structural elements and increased aquiferous system. No similar decrease should be expected for multi-oscular or single-module multi-oscular sponges, where most of the modules are full grown with a relatively low and constant chamber density. For these cases, any reported decrease in the chamber density (and filtration rate), as suggested by [15], would arise if based on the total sponge volume, including an increasing spongocoel volume.

5.2. Closing Remarks: Towards a Better Understanding

From our present examination of F/V versus V, we realized that certain assumptions presented in our recent article on the pumping rate and size of demosponges [16] were not completely correct. Thus, the modeling of a tubular-type demosponge, equivalent to a single-osculum module, was made on the assumption of the constant choanocyte density, which we now find to be unlikely. Furthermore, we realized that the present scaling for (i) single-osculum module ($F/V \sim V^{-1/3}$) cannot be applied to (ii) multi-oscular multi-modular and (iii) single-osculum multi-modular demosponges. We think that the observed and modeled dependence of the filtration rate on the sponge volume for growing single-osculum modules may primarily be governed by a decreasing density of choanocytes with increasing V in all sponge species. However, the hydraulics of the pump and pressure losses of the aquiferous system possibly resulting in a reduction in the choanocyte filtration rate may also play a role, which awaits further assessment. However, the present assessment of F/V versus V among various types of demosponge species should help to clarify the large and confusing variations that we observed, but could not immediately explain.

6. Conclusions

The concept of "modules" was used to classify the demosponges and develop the scaling laws of growth at different stages and the types of sponges. The scaling analysis for single osculum explants leads to a volume-specific filtration rate that scales as $F/V = V^{-1/3}$, which also applies when an aquiferous module grows larger. A multi-oscula sponge is a population of modules each sharing the characteristics of a single-osculum explant. However, many of their modules may have reached their maximal size, and hence their maximal filtration rate, which would imply the scaling $F/V \approx$ constant. A similar scaling would be expected for large pseudo-osculum sponges, provided their volume was taken to be the structural tissue volume that holds the pumping units, and not the total volume that includes the large atrium volume of water. The observed decrease in the F/V ratio by a factor of 5.9 when a very small *Halichondria panicea* explant grows to a near full-grown explant is primarily ascribed to a decrease in the density of the choanocyte chambers.

Author Contributions: H.U.R. and P.S.L. equally contributed with input and text writing. All authors have read and agreed to the published version of the manuscript.

Funding: This research received no external funding.

Institutional Review Board Statement: Not applicable.

Informed Consent Statement: Not applicable.

Data Availability Statement: Not applicable.

Acknowledgments: Thanks are due to Josephine Goldstein for aid with the technical drawing and statistics, and to 4 anonymous reviewers for constructive comments on the manuscript.

Conflicts of Interest: The authors declare no conflict of interest.

References

1. De Voogd, N.J.; Alvarez, B.; Boury-Esnault, N.; Carballo, J.L.; Cárdenas, P.; Díaz, M.-C.; Dohrmann, M.; Downey, R.; Hajdu, E.; Hooper, J.N.A.; et al. World Porifera Database. 2022. Available online: https://www.marinespecies.org/porifera (accessed on 28 March 2022). [CrossRef]
2. Fry, W.G. The sponge as a population: A biometric approach. *Symp. Zool. Soc. Lond.* **1970**, *25*, 135–162.
3. Ereskovskii, A.V. Problems of coloniality, modularity, and individuality in sponges and special features of their mor-phogeneses during growth and asexual reproduction. *Russ. J. Mar. Biol.* **2003**, *29*, 46–56. [CrossRef]
4. Reiswig, H.M. The aquiferous systems of three marine demospongiae. *J. Morph.* **1975**, *145*, 493–502. [CrossRef]
5. Kumala, L.; Riisgård, H.U.; Canfield, D.E. Osculum dynamics and filtration activity studied in small single-osculum explants of the demosponge *Halichondria panicea*. *Mar. Ecol. Prog. Ser.* **2017**, *572*, 117–128. [CrossRef]
6. Goldstein, J.; Riisgård, H.U.; Larsen, P.S. Exhalant jet speed of single-osculum explants of the demosponge *Halichondria panicea* and basic properties of the sponge-pump. *J. Exp. Mar. Biol. Ecol.* **2019**, *511*, 82–90. [CrossRef]
7. Kealy, R.A.; Busk, T.; Goldstein, J.; Larsen, P.S.; Riisgård, H.U. Hydrodynamic characteristics of aquiferous modules in the demosponge *Halichondria panicea*. *Mar. Biol. Res.* **2019**, *15*, 531–540. [CrossRef]
8. Riisgård, H.U.; Thomassen, S.; Jakobsen, H.; Weeks, J.; Larsen, P.S. Suspension feeding in marine sponges *Halichondria panicea* and *Haliclona urceolus*: Effects of temperature on filtration rate and energy cost of pumping. *Mar. Ecol. Prog. Ser.* **1993**, *96*, 177–188. [CrossRef]
9. Larsen, P.S.; Riisgård, H.U. The sponge pump. *J. Theor. Biol.* **1994**, *168*, 53–63. [CrossRef]
10. Thomassen, S.; Riisgård, H.U. Growth and energetics of the sponge *Halichondria panicea*. *Mar. Ecol. Prog. Ser.* **1995**, *128*, 239–246. [CrossRef]
11. McMurray, S.E.; Blum, J.E.; Pawlik, J.R. Redwood of the reef: Growth and age of the giant barrel sponge *Xestospongia muta* in the Floridas Keys. *Mar. Biol.* **2008**, *155*, 159–171. [CrossRef]
12. McMurray, S.E.; Pawlik, J.R.; Finelli, C.M. Trait-mediated ecosystem impacts: How morphology and size affect pumping rates of the Caribbean giant barrel sponge. *Aquat. Biol.* **2014**, *23*, 1–13. [CrossRef]
13. Reiswig, H.M. In situ pumping activities of tropical Demospongiae. *Mar. Biol.* **1971**, *9*, 38–50. [CrossRef]
14. Strehlow, B.W.; Jørgensen, D.; Webster, N.S.; Pineda, M.C.; Duckworth, A. Using a thermistor flowmeter with attached video camera for monitoring sponge excurrent speed and oscular behaviour. *PeerJ* **2016**, *4*, e2761. [CrossRef] [PubMed]
15. Morganti, T.M.; Ribes, M.; Moskovich, R.; Weisz, J.B.; Yahel, G.; Coma, R. In situ pumping rate of 20 marine demosponges is a function of osculum area. *Front. Mar. Sci.* **2021**, *8*, 583188. [CrossRef]

16. Larsen, P.S.; Riisgård, H.U. Pumping rate and size of demosponges-towards an understanding using modeling. *J. Mar. Sci. Eng.* **2021**, *9*, 1308. [CrossRef]
17. R Development Core Team. *R: A Language and Environment for Statistical Computing*; R Foundation for Statistical Computing: Vienna, Austria, 2022. Available online: https://www.R-project.org/ (accessed on 2 March 2022).
18. Fry, W.G. Taxonomy, the individual and the sponge. Biology and Systematics of Colonial organisms. *Syst. Ass. Spec.* **1979**, *11*, 49–80.
19. Pinheiro, U.S.; Hajdu, E. Shallow-water *Aplysina* Nardo (Aplysinidaee Verongida, Demospongiae) from the São Sebastião Channel and its environs (Tropical southwestern Atlantic), with the description of a new species and a literature review of other brazilian records of genus. *Rev. Bras. Zool.* **2001**, *18* (Suppl. S1), 143–160. [CrossRef]
20. Morganti, T.M.; Ribes, M.; Yahel, G.; Coma, R. Size is the major determinant of pumping rates in marine sponges. *Front. Physiol.* **2019**, *10*, 1474. [CrossRef]
21. Reiswig, H.M. Population dynamics of three Jamaican Demospongiae. *Bull. Mar. Sci.* **1973**, *23*, 191–226.
22. Reiswig, H.M. Water transport, respiration and energetics of three tropical marine sponges. *J. Exp. Mar. Biol. Ecol.* **1974**, *14*, 231–249. [CrossRef]
23. Southwell, M.W.; Weisz, J.B.; Martens, C.S.; Lindquist, N. In situ fluxes of dissolved inorganic nitrogen from the sponge community on Conch Reef, Key Largo, Florida. *Limnol. Oceanogr.* **2008**, *53*, 986–996. [CrossRef]
24. Dahihande, A.S.; Thakur, N.L. Temperature- and size-associated differences in the skeletal structures and osculum cross-sectional area influence the pumping rate of contractile sponge *Cinachyrella* cf. *cavernosa*. *Mar. Ecol.* **2019**, *40*, e12565. [CrossRef]
25. Barthel, D. On the ecophysiology of the sponge *Halichondria panicea* in Kiel Bight. I. Substrate specificity, growth and reproduction. *Mar. Ecol. Prog. Ser.* **1986**, *32*, 291–298. [CrossRef]
26. Lüskow, F.; Riisgård, H.U.; Solovyeva, V.; Brewer, J.R. Seasonal changes in bacteria and phytoplankton biomass control the condition index of the demosponge *Halichondria panicea* in temperate Danish waters. *Mar. Ecol. Prog. Ser.* **2019**, *608*, 119–132. [CrossRef]
27. Barthel, D. Influence of different current regimes on the growth form of *Halichondria panicea* Pallas. In *Fossil and Recent Sponges*; Reitner, J., Keupp, H., Eds.; Springer: Berlin/Heidelberg, Germany, 1991; pp. 387–394.
28. Imsiecke, G. Ingestion, digestion, and egestion in *Spongilla lacustris* (Porifera, Spongillidae) after pulse feeding with *Chlamydomonas reinhardtii* (Volvocales). *Zoomorphology* **1993**, *113*, 233–244. [CrossRef]
29. Weissenfels, N. The filtration apparatus for food collection in freshwater sponges (Porifera, Spongillidae). *Zoomorphology* **1992**, *112*, 51–55. [CrossRef]
30. Riisgård, H.U.; Kumala, L.; Charitonidou, K. Using the *F/R*-ratio for an evaluation of the ability of the demosponge *Halichondria panicea* to nourish solely on phytoplankton versus free-living bacteria in the sea. *Mar. Biol. Res.* **2016**, *12*, 907–916. [CrossRef]
31. Ludeman, D.A.; Reidenbach, M.A.; Leys, S.P. The energetic cost of filtration by demosponges and their behavioural response to ambient currents. *J. Exp. Biol.* **2017**, *220*, 995–1007. [CrossRef]
32. Asadzadeh, S.S.; Larsen, P.S.; Riisgård, H.U.; Walther, J.H. Hydrodynamics of the leucon sponge pump. *JRSI* **2019**, *16*, 20180630. [CrossRef]

Journal of
Marine Science and Engineering

MDPI

Article

Actual and Model-Predicted Growth of Sponges—With a Bioenergetic Comparison to Other Filter-Feeders

Hans Ulrik Riisgård [1,*] and Poul S. Larsen [2]

[1] Marine Biological Research Centre, University of Southern Denmark, 5300 Kerteminde, Denmark
[2] Department of Mechanical Engineering, Technical University of Denmark, 2800 Kongens Lyngby, Denmark; psl@mek.dtu.dk
* Correspondence: hur@biology.sdu.dk

Abstract: Sponges are one of the earliest-evolved and simplest groups of animals, but they share basic characteristics with more advanced and later-evolved filter-feeding invertebrates, such as mussels. Sponges are abundant in many coastal regions where they filter large amounts of water for food particles and thus play an important ecological role. Therefore, a better understanding of the bioenergetics and growth of sponges compared to other filter-feeders is important. While the filtration (pumping) rates of many sponge species have been measured as a function of their size, little is known about their rate of growth. Here, we use a bioenergetic growth model for demosponges, based on the energy budget and observations of filtration (F) and respiration rates (R). Because F versus dry weight (W) can be expressed as $F = a_1 W^{b1}$ and the maintenance respiratory rate can be expressed as $R_{\mathrm{m}} = a_2 W^{b2}$, we show that if $b_1 \sim b_2$ the growth rate can be expressed as: $G = aW^{b1}$, and, consequently, the weight-specific growth rate is $\mu = G/W = aW^{b1-1} = aW^b$ where the constant a depends on ambient sponge-available food particles (free-living bacteria and phytoplankton with diameter < ostia diameter). Because the exponent b_1 is close to 1, then $b \sim 0$, which implies $\mu = a$ and thus exponential growth as confirmed in field growth studies. Exponential growth in sponges and in at least some bryozoans is probably unique among filter-feeding invertebrates. Finally, we show that the F/R-ratio and the derived oxygen extraction efficiency in these sponges are similar to other filter-feeding invertebrates, thus reflecting a comparable adaptation to feeding on a thin suspension of bacteria and phytoplankton.

Keywords: bioenergetic growth model; energy budget; filtration rate; respiration; F/R-ratio; filter-feeding

Citation: Riisgård, H.U.; Larsen, P.S. Actual and Model-Predicted Growth of Sponges—With a Bioenergetic Comparison to Other Filter-Feeders. *J. Mar. Sci. Eng.* **2022**, *10*, 607. https://doi.org/10.3390/jmse10050607

Academic Editor: Caterina Longo

Received: 22 March 2022
Accepted: 26 April 2022
Published: 29 April 2022

Publisher's Note: MDPI stays neutral with regard to jurisdictional claims in published maps and institutional affiliations.

1. Introduction

Sponges are simple multicellular filter-feeders that actively pump volumes of water equivalent to five times or more their body volume per minute through their canal system by using flagellated choanocytes, which constitute the pumping and filtering elements of the smallest particles [1–4]. Sponges feed on suspended microscopic particles, including free-living bacteria and phytoplankton [5,6]. Water enters the sponge body through numerous small inhalant openings (ostia) and passes through incurrent canals, where phytoplankton cells with diameters smaller than the ostia diameter but larger than 5 µm are trapped, leading to the choanocyte chambers, where bacteria and other smaller particles are captured by the collar-filter of the choanocytes. Then, the filtered water flows through excurrent canals to be expelled as a jet through an exhalant opening (osculum) [7–9]. The water pumping also ensures ventilation and a supply of oxygen for respiration via diffusive oxygen uptake [10]. Although sponges lack nerves and muscle tissues, coordinated contraction-expansion responses, including partial or complete closure of the osculum to mechanical and chemical stimuli, are common among sponges due to the presence of contractile cells (myocytes) [10–13], which results in temporary reduced or arrested water flow.

Sponges are one of the earliest evolved and simplest groups of animals [14], but they share basic characteristics with more advanced and later-evolved filter-feeding invertebrates such as mussels, in which filter-feeding is a secondary adaptation [15]. Sponges are also abundant today, especially in polar-shelf and tropical-reef communities as well as in many coastal regions, where they filter large amounts of water for food particles and thus play an important ecological role [5,16–21]. Therefore, a better understanding of the bioenergetics and growth of sponges in comparison with other filter-feeders is important.

Here, we first use an earlier approach for setting up a bioenergetic growth model, based on the energy budget and observations of filtration and respiration rates, which suggest that the growth of sponges is exponential; next, we use data in the literature to verify this hypothesis. Finally, we compare sponges to other filter-feeding invertebrates in order to compare the evolutionary adaptation of these animals to feed on the same thin soup of bacteria and phytoplankton.

2. Materials and Methods

Based on sponge data in the literature, we first set up a growth model based on the energy budget for growth (G) by making use of near identical exponents in the power functions for filtration rate (F) and respiration rate (R) versus body dry weight W. This development follows the earlier approach for the growth of the blue mussel *Mytilus edulis* [22], for which F or $R = aW^{0.66}$ leads to a decrease in the weight-specific growth rate of mussels with increasing size, given by: $\mu = G/W = aW^{0.66} - 1 = aW^{-0.34}$, which showed good agreement with field data. For many sponges, it is suggested that exponents b_1 and b_2 of power functions for F and R are close to $b_1 \sim b_2 \sim 1$, and, therefore, the model predicts that the weight-specific growth rate must be constant with increasing sponge size, which implies that the growth must be exponential. This hypothesis is subsequently verified by sponge-growth data from the literature obtained in field experiments conducted in periods with positive growth, typically in the spring. The development of sponge size in time intervals was used to estimate the specific growth rates in each interval and was subsequently used for the evaluation of growth patterns. Exponential growth is characterized by a constant specific growth rate, which is reflected as a trendless scatter of the interval-specific growth rates, in contrast to a power function growth pattern, where the interval-specific growth rate will decrease with increasing size; for such data, an exponential curve fit will systematically underestimate the actual data in the first half period and then overestimate the data in the remaining period. Data were replotted from other publications using an in-house graphical program, 'Gtpoints', which generates a table of data point coordinates according to the scales of axes in a *.bmp image of a given graph.

Exponential regression curves (LM) were fitted in [23] for growth rate estimates based on wet weight, ash-free dry weight, or length of sponge over time.

3. Results and Discussion

3.1. Sponge Growth Model and Test of Exponents

The growth of a sponge can be expressed by the energy (or carbon) budget as:

$$G = I - R - E = A - R \tag{1}$$

where G = growth (production), I = ingestion, R = respiration (total) = R_m (maintenance respiration) + R_g (growth respiration, i.e., metabolic cost of synthesizing new biomass), E = excretion, and A = assimilated food. The budget can also be written as $G = (F \times C \times AE) - (R_m + R_g)$, where F = filtration rate, C = food concentration, and $AE = A/I$ = assimilation efficiency. Thus, equating the rate of the net intake of nutritional energy to the sum of various rates of consumption, the energy balance for a growing sponge may now be written as:

$$G = [(F \times C \times AE) - R_m]/a_0 \tag{2}$$

where the constant a_0 is the metabolic cost of growth, which constitutes a certain amount of energy equivalent to a constant percentage of the growth (biomass production). Because the filtration rate (F) of a sponge can be estimated from the sponge dry weight (W) according to $F = a_1 W^{b1}$, and the maintenance respiratory rate (R_m) can be estimated according to $R_m = a_2 W^{b2}$, then if $b_1 \approx b_2$, the growth rate may now be expressed as:

$$G = (C \times AE \times a_1 - a_2)\, W^{b1}/a_0 = aW^{b1} \tag{3}$$

which seems to be an equation that would apply to sponges in general (and other marine filter-feeding invertebrates, see later). Thus, [24] found that $b_1 = 0.914$ and $b_2 = 0.927$ for *Halichondria panicea*, while [17] found that $b_1 \approx b_2 \approx 1$ for three tropical marine sponges (up to a size of 2.5 l sponge). Because the percentage of oxygen removed from the water pumped through the sponge is rather constant [17], this implies that b_2 tends to be similar to b_1, as in a recent review of the volume-specific pumping rate of demosponges versus sponge volume (V) approximated by the power-law $F/V = a_3 V^{b3}$ [25], where $b_3 = b_1 - 1$, assuming that dry weight was proportional to volume. By comparing the exponents reported by various authors, it appears that $b_1 \approx 1$ or $b_1 \geq 0.9$ [17,24,26–32], but in other cases, $b_1 < 0.9$ has been reported [21].

Therefore, if $b_1 \approx 1$ in Equation (3), the resulting model for weight-specific growth rate ($\mu = G/W = aW^{b1}/W$) becomes:

$$\mu = a \tag{4}$$

which is a constant and thus the growth is exponential.

On the other hand, if $b_1 \approx b_2 \approx 0.9$, then $b = 0.9 - 1 = -0.1$, and the resulting model may be expressed as:

$$\mu = aW^{-0.1} \tag{5}$$

Here, using published growth data, we first test the prediction of Equation (4), which, with Equation (3), may be expressed as:

$$\mu = G/W = (C \times AE \times F - R)/a_0 \tag{6}$$

In the case of *Halichondria panicea*, the following numbers apply: F = volume specific filtration rate = 6.1 mL water (ml sponge)$^{-1}$ min^{-1} [4], R = weight specific respiration rate = 7.93 μM O$_2$ h^{-1} (g W)$^{-1}$ ([33] = (7.93 × 32/1000 × 0.7) = 0.178 mL O$_2$ h^{-1} (g W)$^{-1}$, and $a_0 = 2.39$ since the cost of growth (SDA) is equivalent to 139% of the biomass production [24], where W henceforth denotes sponge dry weight. Furthermore, 1 mL sponge = 90.019 mg W [6], 1 mL O$_2$ = 0.46 mg C [34], and 1 mg W = 0.142 g C [24]. Assuming $AE = 0.8$ and inserting numbers in Equation (6), we find: $\mu(\mathrm{d}^{-1}) = G/W = [[C \times (\mu\mathrm{g\ C\ L^{-1}}) \times 0.8 \times 6.1 \times 60 \times 24/1000$ (per mL sponge = 12.78 mg C) $- 0.178 \times 0.46 \times 24$ (per g W = 0.142 g C)]/2.39]/1000, or:

$$\mu = a = [(C \times 0.55 - 13.8)/2.39]/1000 \tag{7}$$

From this, for $\mu = 0$, the maintenance food concentration is estimated at $C_m = (13.8/0.55 =)$ 25.1 μg C L^{-1}. If the total sponge-available carbon biomass (TCB, i.e., free-living bacteria and phytoplankton with diameters smaller than the ostia diameter) is 5 times as large (i.e., $C = 125.5$ μg C L^{-1}), the predicted specific growth rate is estimated as $\mu = [(125.5 \times 0.55 - 13.8)/2.39 = 23.11$ μg C d^{-1}/1000 μg C =] 0.023 d^{-1} = 2.3% d^{-1}, which may be compared with actually measured growth rates in the field as appears from the following examples. It should be mentioned that the influence of temperature, salinity, and spawning have not been addressed in Equation (7) and that $AE = 0.8$ will decrease if the ingestion exceeds the amount of food needed for maximum (biologically possible) growth.

3.2. Verification of Hypothesis: Growth Rates of Sponges in the Field

Example 1. *The growth of Halichondria panicea was measured in the inlet to Kerteminde Fjord, Denmark, in a 104-d experiment conducted by [6]; see Figure 1. The exponential curve fit indicates a weight-specific growth rate of $\mu = 0.6\%\ d^{-1}$ (Figure 1), which may be compared to the algebraic mean $\mu = 0.62\%\ d^{-1}$ (Table A1), but the big scatter in calculated μ-values during the time of growth indicates no meaningful correlation between μ and sponge dry weight. The growth experiment was conducted in the period April-August, when the temperature was 20 °C at both the beginning and end of the experiment, with a peak of about 25 °C between June and July. The growth rate of 0.6% d^{-1} indicates, according to Equation (7), that the mean food availability (TCB) had been about 50 μg C L^{-1}, which may be compared to the mean TCB of about 90 μg C L^{-1} measured in the ambient water during the sponge-growth period March–August [6].*

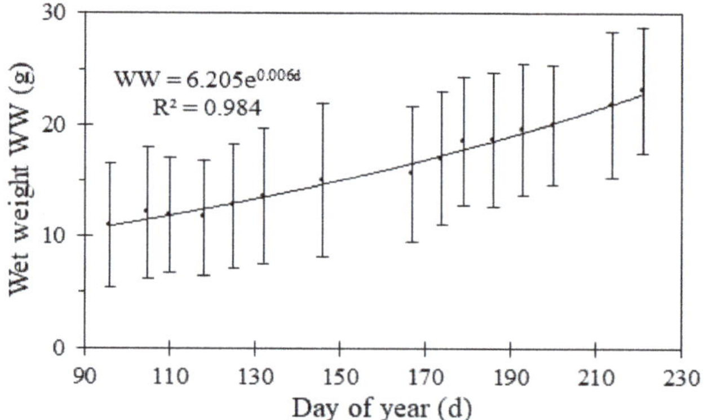

Figure 1. *Halichondria panicea.* Development of mean ± SD wet weight as a function day of year. Exponential regression curve (LM, $t_{0.015,\ 13} = 9.3 \times 10^{-5}$, $p = 6.8 \times 10^{-13}$) shows a weight-specific growth rate of 0.6% d^{-1}. Data from [6].

Example 2. *Growth of Halichondria panicea in the field at Kiel Bight was measured by [35,36]. Replotting the data of [35] for the period March-June of intense growth shows an approximately exponential growth AFDW = 83.6 $e^{0.007d}$ (Figure 2), which suggests an average specific growth rate of $\mu_{avg} \sim 0.7\%\ d^{-1}$. Calculated values of specific growth rate μ show a large scatter but the same value of an algebraic average growth rate. Temperature seemed to be the controlling factor for growth during this period, where it was recorded to increase from 1.8 to 13.8 °C (Table A2), but the concurrent increase in phytoplankton and bacteria biomass was not monitored. Therefore, both biological and physical effects might have contributed to the accelerated growth during the latter part of the period. Later, [36] presents data on biomass changes in a field study on populations of H. panicea at 3 different water depths of 6, 8, and 10 m. Replotting the data for the growth period up to the peak values in August shows growth in terms of sponge biomass (ashfree dry mass) AFDM g m^{-2}, with exponential growth constants being $\mu_{avg} = 1.0, 2.2$ and $0.6\%\ d^{-1}$, respectively (Figure 3). The temperature at 10 m depth was observed to increase from 1.4 to 16.1 °C during the period of growth from February to August of 1984.*

Example 3. *The growth in terms of body length of two Indo-Pacific sponges, Neopetrosia sp. and Stylissa massa, were measured by [37] from late November to March of 2003 for various farming conditions. While S. massa showed low or no growth, it was significant for Neopetrosia sp. for most treatments, reaching an exponential growth constant in terms of length L of specimens that we have derived from a replot of [37] to be $\mu_{L,avg} = 0.7\%\ d^{-1}$ (Figure 4). To estimate an equivalent exponent for this growth in terms of volume we use the data from [37] for initial (i) and end (e) values of length and volume of Neopetrosia sp. and assume the relation $V \sim L^n$. This leads to the value $n = ln(V_e/V_i)/ln(L_e/L_i) = ln(48.3/10.8)/ln(10.8/5.9) = 2.48$, and $\mu_{V,avg} = 2.48 \times 0.71 = 1.76\%\ d^{-1}$.*

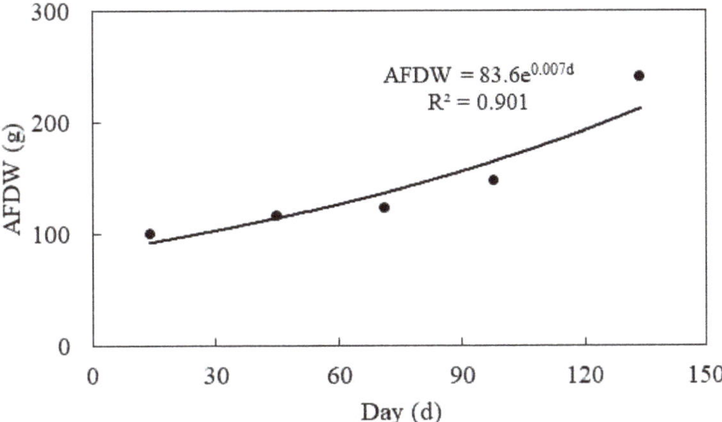

Figure 2. *Halichondria panicea.* Exponential regression curve (LM, $t_{0.045, 3} = 5.8 \times 10^{-4}$, $p = 0.014$) showing an approximately average growth rate of 0.7% d^{-1}. Based on data from [35] shown in Table A2.

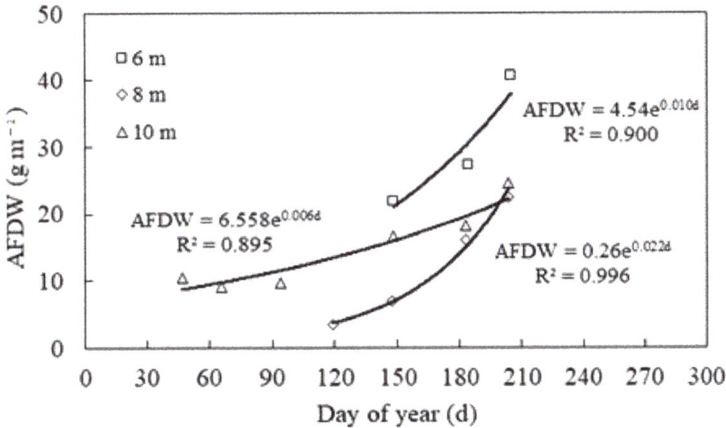

Figure 3. *Halichondria panicea.* Exponential regression curves at 3 water depths (6 m: LM, $t_{0.006, 1} = 1.5 \times 10^{-3}$, $p = 0.205$; 8 m: LM, $t_{0.030, 2} = 4.6 \times 10^{-4}$, $p = 0.002$; 10 m: LM, $t_{0.064, 4} = 4.5 \times 10^{-4}$, $p = 0.004$) showing approximate average growth rates of $\mu_{avg} = 1.0$, 2.2 and 0.6% d^{-1}. Based on data from [36] shown in Table A3.

Example 4. *The growth rate of the demosponge* Haliclona oculata *was studied in its natural environment, Oosterschelde, in the Netherlands, by* [38] *who assumed "exponential growth" and found that the maximum average ($\pm$SD) volume-specific growth rate for 11 monitored specimens was 1.18 $\pm$ 0.35% d^{-1} in the beginning of May 2006.*

From the above examples, it appears that sponge growth in the field is approximately exponential and that the weight-specific growth rate is constant, typically a few % d^{-1} or less. In laboratory and field experiments conducted by [24] with *Halichondria panicea* the maximum measured growth rate was about 4% d^{-1}, and [38,39] give data on wet weight-based exponential growth of the freshwater sponge *Spongilla lacustris*, indicating an exponent of $\mu_{WW} \sim 4.5$% d^{-1} for dark aposymbiotic conditions. Thus, the maximum possible growth rate of sponges seems to be about 4% d^{-1} and in case of *H. panicea* maximum growth may take place at about 8 times the TCB maintenance concentration of C_m = 25.1 µg C L^{-1}. At higher TCB concentrations, the AE = 0.8 in Equation (7) will decrease, and no further increase in growth can be expected. The specific growth rate linearly increases from C_m to C_{max} (see also Figure A1).

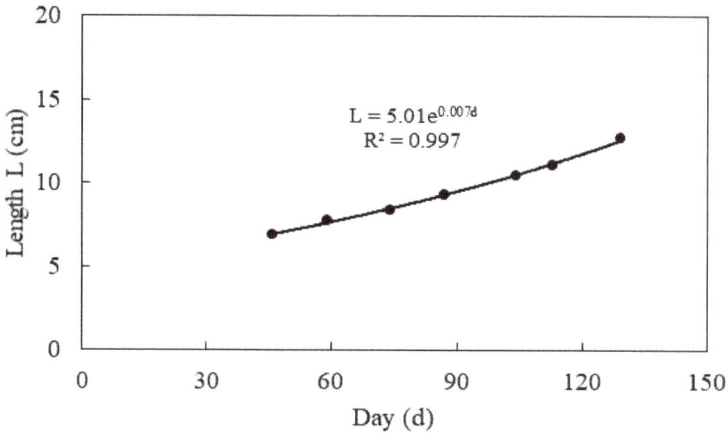

Figure 4. *Neopetrosia* sp. Exponential regression curve (LM, $t_{0.006, 5}$ = 7.9 × 10^{-5}, p = 2.0 × 10^{-7}) confirming an essentially constant average growth rate of 0.7% d^{-1} in terms of specimen length L. Based on data from [37] shown in Table A4.

3.3. Evolutionary Adaptation

In the following 3 sections, we compare sponges to other filter-feeding invertebrates in order to compare the evolutionary adaptation of these animals to feeding on the same thin soup of bacteria and phytoplankton. In order to understand how sponges comply with the performance requirements for being a true filter-feeder, we first compare the *F/R*-ratio (amount of water filtered per ml of oxygen consumed). Next, based on data from the literature on the *F/R*-ratio in various sponge species, we estimate the oxygen-extraction efficiency to evaluate to what extend the respiration rate may be dependent on the filtration rate. Finally, we discuss how sponges, like other filter-feeders in temperate waters, cope with low phytoplankton concentrations during winter.

3.3.1. *F/R*-Ratio

The *F/R*-ratio (liters of water filtered per ml of oxygen respired) can be determined using the above given volume specific rates for *Halichondria panicea*, F = [6.1 × 60/1000 = 0.366 L h^{-1} (mL sponge)$^{-1}$, or = 0.366/90.019 = 4.07 × 10^{-3} L h^{-1}] (mg dry weight sponge)$^{-1}$ and R = 0.178 × 10^{-3} mL O$_2$ h^{-1} (mg dry weight sponge)$^{-1}$ as: F/R = (4.07/0.178 = 22.9 L) water filtered per mL O$_2$ consumed. This *F/R*-ratio is well above the minimum reference value of 10 L water (mL O$_2$)$^{-1}$ for a true marine filter-feeding invertebrate [40], and to balance the sponge's energy requirements, the particulate organic carbon (sponge-available phytoplankton and free-living bacteria) should be $>C_m$ = 25.1 µg C L^{-1}. [30] shows the measured *F/R*-ratio in the demosponge *Callyspongia vaginalis* to be approximately 0.42 ± 0.03 L water (µmol O$_2$)$^{-1}$, which converts to F/R = (420 L/22.4 mL O$_2$) = 18.8 L water (mL O$_2$)$^{-1}$, thus comparing well with the above example of *H. panicea*. As mentioned earlier, [17] found that $b_1 \approx b_2 \approx 1$ for three tropical marine sponges, but an actual growth rate was not reported. However, the F/R-ratios were reported to be: *Mycale* sp. = 22.8 L water (mL O$_2$)$^{-1}$; Tethya crypta = 19.6 L water (mL O$_2$)$^{-1}$; and *Verongia giganta* = 4.1 L water (mL O$_2$)$^{-1}$. The first two species comply with the performance requirement for being a true filter-feeder, whereas *V. giganta* does not, because it "consists of a tripartite community: sponge-bacteria-polychaete" [17]. We think that the *F/R*-ratio is a reliable check in support of $b \approx 0$ in *Mycale* sp. and *T. crypta*. [30] shows for 5 demosponges that the weight-specific oxygen consumption (R) versus weight-specific filtration rate (F) could be described as $R = a_4 F^{b4}$ where $b_4 \approx 1$ (or 0.9416), which supports the idea that $b_1 \approx b_2 \approx 1$ in these sponges.

3.3.2. Oxygen Uptake and Extraction Efficiency

When the F/R-ratio is known, the oxygen extraction efficiency may be estimated as the reciprocal of the total amount of oxygen passing through the sponge per ml of oxygen taken up by the sponge. The following conversion factors may be used: 1 mg O_2 = 0.7 mL O_2; in fully oxygenated seawater there is 9 mg O_2 L^{-1} = 6.3 mL O_2 L^{-1} available, so for 22.9 L there would be 22.9 × 6.3 = 144.3 mL O_2 available. But the uptake was only 1 mL O_2 for 22.9 L water pumped according to the estimated F/R-ratio for *Halichondria panicea*. Therefore, the extraction efficiency in this case was EE = 1/144.3 = 0.007, or 0.7%. This is in agreement with [15], who found that the oxygen extraction efficiency is 1% or less in coastal filter feeders. Like in other filter-feeding animals such as mussels, the ventilatory currents are laminar in sponges, and oxygen in the water is only taken up by diffusion, which implies that only a small fraction of the oxygen in the water pumped through the sponge is available for respiration. In *Verongia giganta*, where F/R = 4.1 L water (mL $O_2)^{-1}$ (see previous section), the extraction efficiency is calculated at EE = (1 mL O_2/(4.1 × 6.3) mL O_2 = 3.9%), which is 5 times higher than for *Mycale* sp., *Tethya crypta*, and the above example with *H. panicea*.

Reduced flow due to increased pressure losses in canal systems [25] or to closure of the osculum will result in low and high values of F/R and EE, respectively, as observed for example for *Verongia giganta*. As shown for the filter-feeding blue mussel *Mytilus edulis* [15], the respiration rate in a sponge is probably independent of filtration rate above about 20% or less of the filtration rate capacity of a sponge because the extraction efficiency increases with a decreasing filtration rate. Like other filter-feeders, sponges also experience low phytoplankton concentrations in temperate waters in the northern hemisphere during winter [6]. The lower threshold of total available carbon concentrations (phytoplankton plus bacterial carbon) covering the maintenance cost of *H. panicea* is found here (Example 1) to be around 50 µg C L^{-1}. *M. edulis* copes with low phytoplankton content during winter by reducing its valve gape during starvation periods [41]. A study by [42] demonstrated that alternating closing-opening of valves causes a simultaneous strong decrease in oxygen concentration in the mantle cavity and thus a reduction of the respiration rate, and in this way *M. edulis* saves energy during starvation periods, a statement supported by a starvation experiment where the metabolic weight loss was reduced 12.3 times during 159 days of starvation [41]. It is well-known that the modular colonial *H. panicea*, with an osculum on each module, is able to close its oscula and thus reduce or stop the water-through flow [4,43,44], which may result in reduced respiration and eventually in internal anoxia [45], and, therefore, closing-opening of the oscula during starvation periods might theoretically be an energy-saving mechanism comparable to that found in *M. edulis*. However, our preliminary observations on *H. panicea* do not support the existence of such a mechanism.

3.3.3. Growth and Respiration

The specific respiration rate in response to growth (specific dynamic action, SDA) constitutes 139% of the biomass production in *Halichondria panicea* [24]. This percentage, used in the present growth model (see Equation (7)), makes up a very substantial proportion of the total energy released by respiration compared to other filter-feeding invertebrates, where the cost of growth is typically between 12 and 20% of the biomass production [46]. Due to their simple structure, sponges may be regarded as colonies composed of water-pumping choanocytes that are structurally and functionally near identical to free-living choanoflagellates in the sea [47]. In these organisms, with which choanocytes share properties, energy used for maintenance only constitutes a small fraction of the energy required for growth. Thus, a doubling of the specific growth rate of e.g., a flagellate protozoan may result in a doubling of the specific respiration rate, indicating that the energy cost of growth (mainly macromolecular synthesis) is equivalent to that of the actual growth [46].

As for the maintenance respiration, the total respiration (R) as a function of body dry weight (W) is usually described by the power function $R = a_5 W^{b5}$ where the b_5-exponent is frequently close to 0.75. However, the "3/4 power scaling" is not a 'natural constant' because many exponents differ from $b_5 = 0.75$ [46]. Thus, $b_5 = 0.66$ in the blue mussel *Mytilus edulis* [48], $b_5 = 0.68$ in the ascidian *Ciona intestinalis* [49], $b_5 = 0.86$ in the jellyfish *Aurelia aurita* [50], $b_5 = 1.2$ in the facultative filter-feeding polychate *Nereis diversicolor* [51], $b_5 = 0.93$ in the demosponge *Halichondria panicea* [24], and $b_5 = 1$ in three tropical sponges [17]. A bioenergetic growth model based on the energy budget and making use of near identical exponents in the power functions for filtration rate (F) and respiration rate (R) versus W (i.e. F or $R = a_2 W^{b2}$) has earlier been developed for the blue mussel *Mytilus edulis*, where $b_1 \approx b_2 = 0.66$ and where it was found that actual growth rates in the field in general were in good agreement with the model and—as predicted—that the weight-specific growth rate decreased with body dry weight as $\mu = aW^{0.66-1} = aW^{-0.34}$ [22]. As a consequence, the weight specific growth rate of 7.8% d^{-1} for a 0.01 g dry weight *M. edulis* exposed to 3 µg chl a L^{-1} gradually decreases to 1.6% d^{-1} for a 1 g mussel, thus showing that the growth is not exponential. The constant specific growth rate with increasing size and thus exponential growth in sponges (and some bryozoans, see later) is unique and does not exist in other filter-feeding invertebrates where $b_1 \approx b_2 < 1$.

Referring to the "general" model for metabolic scaling $R = aW^{0.75}$, the mass-specific metabolic scaling becomes $R/W = aW^{-0.25}$, which exponent $b = -0.25$ [20] is found (apparently suggesting that F/V follows R/W) to be "consistent with the measured exponent for three of five species" of sponges in which they had measured exponents for volume-specific filtration rate (F/V) versus sponge volume (V): $F/V = a_3 V^{b3}$ and found $b_3 = -0.19, -0.20, -0.22, -0.49,$ and -0.70 for the five sponge species, respectively [20]. However, such comparison with a suggested "general" metabolic scaling is unwarranted, and the negative b_3-exponents may need another explanation (see later).

The volume of a sponge is not always closely related to the living sponge biomass, which is evident from the seasonal variation in the sponge condition index CI = ratio of organic to inorganic matter = AFDW/(DW − AFDW) [35,52]. Thus, a decreasing value of CI reflects a decreasing relative density of water-pumping choanocytes in a sponge, hence a lower pumping rate for a given sponge volume. So, if CI decreases while volume increases, this may explain the negative exponents for volume-specific filtration rate versus sponge volume in these sponges. Thus, it can be put forward as a hypothesis that spicules with decreasing CI form an increasing and major component of the volume in some sponge species, and thus a decreasing volume-specific filtration rate is associated with increasing size; see also [53]. In addition, or alternatively, the observed dependence of filtration rate on size of sponges "might primarily be governed by the hydraulics of pump and pressure losses of the aquiferous system" and not by, e.g., "a reducing density of choanocytes with increasing size", as suggested by [25]; see next section. From the present study, it is obvious that sponges have many features in common with other filter-feeding invertebrates. Thus, the F/R-ratios and oxygen extraction efficiencies are comparable because all filter-feeding organisms have to cope with the same challenge of living in a thin soup of suspended microscopic food particles.

Sponges are modular organisms that consist of a set of repetitive modules. Likewise, filter-feeding bryozoans are colonial animals that consist of a set of repetitive zooids, which may also give rise to exponential growth, e.g., in *Celleporella hyalina, Electra pilosa* [54,55], *Cryptosula pallasiana*, and other bryozoan species [56]. Because the individual filtration rate and respiration rate of a module, or of a zooid, remain the same when a sponge or bryozoan colony grows, both the total filtration rate (F) and respiration rate (R) of the organisms increase linearly with the increasing number of modules/zooids (W), i.e., F and $R = a_2 W$, which implies exponential growth according to the bioenergetics growth model. However, in many bryozoan species the rate of asexual zooid replication increases with colony size [57], and, therefore, the rate of growth in the number of zooids occurs in a different way, following power function. Exponential growth probably does not exist in non-modular and non-colonial filter-feeding invertebrates where the exponents in the equation for F and $R = a_2 W^{b2}$ are usually <1 and tend to be equal. In such cases where $b_2 < 1$, the growth follows a power function. Thus, for $b_2 = -0.34$ in the blue mussel *Mytilus edulis* the weight-specific growth rate as a function of W can be described by a power regression line in a log-log plot in which the slope has been found to be close to the predicted $b_2 = -0.34$ [21]. In the filter-feeding jellyfish *Aurelia aurita* it has been shown that $b_2 = -0.2$, which shows that the weight-specific growth rate is not constant but decreasing with size as reflected in systematic deviation between exponential regression curve fit for W versus time that underestimates W in the first half of the growth period while it overestimates it in the second period [58]. Thus, although growth versus time may be fitted approximately by an exponential curve a systematic deviation indicates that the specific growth rate is not constant. No such systematic deviations have been observed in the present study, which supports that the growth of sponges is exponential, as does the bioenergetic model and the application of the concept of modules.

An explanation for this may—as a theory—be derived from the high F/R-ratio and the low oxygen extraction efficiency in filter-feeding invertebrates. Because the proportion of biomass with low metabolism (e.g., lipid and glycogen store, gonads) may increase with body size, the weight-specific respiration (R/W) may concurrently decrease due to a negative exponent ($b_2 - 1$) in the equation $R/W = a_2 W^{b2-1}$. When the biomass of a filter-feeder increases, the total respiration consequently increases, but the oxygen demand should easily be met by an increase in the oxygen extraction efficiency. However, the animal must also increase the filtration rate, and thus the ingestion of food needed to cover the respiratory need to ensure that the F/R-ratio remains unchanged because a reduction in the F/R-ratio will cause starvation. Thus, the exponents in the equation for F and $R = a_2 W^{b2}$ may have (during the evolution) become near equal depending on species and adaptation to living site. Due to the simple structure of sponges, which have no organs or real tissue that may store energy reserves (such as fat, lipids, or glycogen) [6,35] to overcome starvation periods, the exponents for F and R versus W tend to be close to 1, as seen in those sponge species where both F and R have been measured, supported by growth experiments and model predictions presented in this study.

4. Conclusions

The power function exponents $b_1 \sim b_2 \sim 1$ for F and R versus W may probably apply to most demosponges, and therefore Equation (3) may be a general sponge equation. The resulting model for the weight-specific growth rate is a constant, and the growth is therefore exponential. This prediction is confirmed by actual field growth data for a group of sponges for which the F/R-ratios and oxygen extraction efficiencies are comparable to the values of other filter feeders. However, the constant specific growth rate with increasing size in sponges and some bryozoans is unique, and exponential growth probably does not exist in other filter-feeding invertebrates where $b_1 \approx b_2 < 1$, giving rise to a decreasing weight-specific growth rate with increasing body size.

Author Contributions: H.U.R. and P.S.L. equally contributed with input and text writing. All authors have read and agreed to the published version of the manuscript.

Funding: This research received no external funding.

Institutional Review Board Statement: Not applicable.

Informed Consent Statement: Not applicable.

Data Availability Statement: Not applicable.

Acknowledgments: Thanks are due to Josephine Goldstein for help with statistics.

Conflicts of Interest: The authors declare no conflict of interest.

Appendix A

Table A1. *Halichondria panicea*. Development of sponge wet weight (*WW*), dry weight (*W*), average dry weight (W_{avg}) in time interval and weight-specific growth rate (μ) as a function of time (d) in growth experiment with sponge explants in the inlet to Kerteminde Fjord, Denmark. Data used for calculation of μ are from [6].

Time (d)	Day of Year	WW (g)	W (mg)	W_{avg} (mg)	μ (% d^{-1})
0	96	11.0	989		
9	105	12.2	1094	1040	
14	110	11.9	1073	1083	0.45
22	118	11.7	1052	1062	−0.39
29	125	12.7	1146	1098	0.41
36	132	13.6	1224	1184	1.08
50	146	15.0	1353	1287	1.19
71	167	15.6	1404	1378	0.49
78	174	17.0	1530	1466	0.29
83	179	18.6	1672	1599	1.25
90	186	18.7	1679	1675	0.93
97	193	19.5	1759	1719	0.36
104	200	20.0	1796	1777	0.48
118	214	21.8	1966	1879	0.79
125	221	23.1	2078	2021	0.52
				Mean	0.62

Table A2. *Halichondria panicea*. Development of sponge biomass (AFDW) and temperature obtained by replotting data from [34] based on a field experiment at Boknis Eck, Western Baltic Sea, at 10 m and used here for calculation of the weight-specific growth rate (μ).

Time (d)	AFDW (g)	AFDW$_{avg}$ (g)	μ (% d^{-1})	T (°C)
14	99.8			1.8
45	115.8	107.5	0.480	3.0
71	123.2	119.5	0.236	4.7
98	147.9	135.0	0.664	7.5
134	241.0	188.8	1.365	13.8
		Average	0.69	7.26

Table A3. *Halichondria panicea*. Development of biomass (AFDW) in natural sponge populations at Boknis Eck, Western Baltic Sea, at 3 different water depths of 6, 8 and 10 m obtained by replotting data from [35]. Day zero = 1 January 1984.

6 m depth			
Day of 1984 (d)	AFDW (g m^{-2})	AFDW$_{avg}$ (g m^{-2})	μ (% d^{-1})
148	21.9		
184	27.2	24.4	0.61
205	40.7	33.3	1.85
		Average	1.23
8 m depth			
119	3.5		

Table A3. *Cont.*

147	6.9	4.95	2.41
183	16.0	10.54	2.34
204	22.5	18.99	1.58
		Average	2.11
10 m depth			
47	10.4	4.32	2.54
66	9.1	9.71	0.20
94	9.6	9.31	
148	16.6	12.59	1.03
183	18.2	17.35	0.27
204	24.5	21.13	1.41
		Average	1.09

Table A4. *Neopetrosia* sp. Growth data in terms of specimen length obtained by replotting data from [36].

Time	Length	L_{avg}	μ_L
(d)	(cm)	(cm)	(% d^{-1})
46	6.9		
59	7.8	7.32	0.86
74	8.4	8.07	0.51
87	9.3	8.83	0.78
104	10.5	9.89	0.75
113	11.1	10.81	0.58
129	12.7	11.86	0.84
		Average	0.72

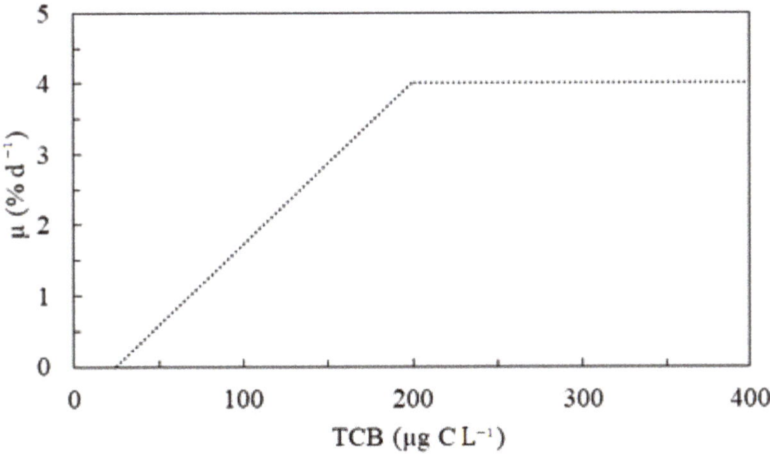

Figure A1. *Halichondria panicea.* Estimated weight-specific growth rate as a function of TCB (total sponge-available carbon biomass) according to Equation (7) and suggested maximum possible growth rate $\mu = 4\%$ d^{-1} at $C_{max} = 200$ µg C L^{-1} = 8 × C_m (maintenance TCB). The increase in specific growth rate is linear between C_m and C_{max}.

References

1. Larsen, P.S.; Riisgård, H.U. The Sponge Pump. *J. Theor. Biol.* **1994**, *168*, 53–63. [CrossRef]
2. De Goeij, J.M.; van den Berg, H.; van Oostveen, M.M.; Epping, E.H.; Van Duyl, F.C. Major bulk dis-solved organic carbon (DOC) removal by encrusting coral reef cavity sponges. *Mar. Ecol. Prog. Ser.* **2008**, *357*, 139–151. [CrossRef]

3. Leys, S.P.; Yahel, G.; Reidenbach, M.A.; Tunnicliffe, V.; Shavit, U.; Reiswig, H.M. The Sponge Pump: The Role of Current Induced Flow in the Design of the Sponge Body Plan. *PLoS ONE* **2011**, *6*, e27787. [CrossRef] [PubMed]

4. Riisgård, H.U.; Kumala, L.; Charitonidou, K. Using the F/R-ratio for an evaluation of the ability of the demosponge *Halichondria panicea* to nourish solely on phytoplankton versus free-living bacteria in the sea. *Mar. Biol. Res.* **2016**, *12*, 907–916. [CrossRef]

5. Reiswig, H.M. In situ pumping activities of tropical Demospongiae. *Mar. Biol.* **1971**, *9*, 38–50. [CrossRef]

6. Lüskow, F.; Riisgård, H.U.; Solovyeva, V.; Brewer, J.R. Seasonal changes in bacteria and phytoplankton biomass control the condition index of the demosponge *Halichondria panicea* in temperate Danish waters. *Mar. Ecol. Prog. Ser.* **2019**, *608*, 119–132. [CrossRef]

7. Weissenfels, N. The filtration apparatus for food collection in freshwater sponges (Porifera, Spongillidae). *Zoomorphology* **1992**, *112*, 51–55. [CrossRef]

8. Imsiecke, G. Ingestion, digestion, and egestion in *Spongilla lacustris* (Porifera, Spongillidae) after pulse feeding with *Chlamydomonas reinhardtii* (Volvocales). *Zoomorphology* **1993**, *113*, 233–244. [CrossRef]

9. Osinga, R.; Tramper, J.; Wijffels, R.H. Cultivation of Marine Sponges. *Mar. Biotechnol.* **1999**, *1*, 509–532. [CrossRef]

10. Bergquist, P.R. *Sponges*; University of California Press: Berkeley, CA, USA, 1978.

11. Bagby, R.M. The fine structure of myocytes in the sponges *Microciona prolifera* (Ellis and So-lander) and *Tedania ignis* (Duchassaing and Michelotti). *J. Morphol.* **1966**, *118*, 167–181. [CrossRef]

12. Elliott, G.R.D.; Leys, S.P. Coordinated contractions effectively expel water from aquiferous system of a freshwater sponge. *J. Exp. Biol.* **2007**, *210*, 3736–3748. [CrossRef] [PubMed]

13. Ellwanger, K.; Eich, A.; Nickel, M. GABA and glutamate specifically induce contractions in the sponge *Tethya wilhelma*. *J. Comp. Physiol. A Sens. Neural Behav. Physiol.* **2007**, *193*, 1–11. [CrossRef] [PubMed]

14. Nielsen, C. Six major steps in animal evolution: Are we dervied sponge larvae? *Evol. Dev.* **2008**, *10*, 241–257. [CrossRef] [PubMed]

15. Jørgensen, C.B.; Møhlenberg, F.; Sten-Knudsen, O. Nature of relation between ventilation and oxygen consumption in filter feeders. *Mar. Ecol. Prog. Ser.* **1986**, *29*, 73–88. [CrossRef]

16. Reiswig, H.M. Population dynamics of three Jamaican Demospongiae. *Bull. Mar. Sci.* **1973**, *23*, 191–226.

17. Reiswig, H.M. Water transport, respiration and energetics of three tropical marine sponges. *J. Exp. Mar. Biol. Ecol.* **1974**, *14*, 231–249. [CrossRef]

18. Pile, A.J.; Patterson, M.R.; Witman, J.D. In situ grazing on plankton < 10 μm by the boreal sponge Mycale lingua. *Mar. Ecol.* **1996**, *141*, 95–102.

19. Ribes, M.; Coma, R.; Gili, J. Natural diet and grazing rate of the temperate sponge *Dysidea avara* (Demospongiae, Dendroceratida) throughout an annual cycle. *Mar. Ecol. Prog. Ser.* **1999**, *176*, 179–190. [CrossRef]

20. Morganti, T.; Ribes, M.; Yahel, G.; Coma, R. Size Is the Major Determinant of Pumping Rates in Marine Sponges. *Front. Physiol.* **2019**, *10*, 1474. [CrossRef]

21. Morganti, T.M.; Ribes, M.; Moskovich, R.; Weisz, J.B.; Yahel, G.; Coma, R. In situ pumping rate of 20 marine demosponges is a function of osculum area. *Front. Mar. Sci.* **2021**, *8*, 583188. [CrossRef]

22. Riisgård, H.U.; Lundgreen, K.; Larsen, P.S. Potential for production of 'mini-mussels' in Great Belt (Denmark) evaluated on basis of actual and modeled growth of young mussels Mytilus edulis. *Aquac. Int.* **2014**, *22*, 859–885. [CrossRef]

23. R Development Core Team R. A Language and Environment for Statistical Computing; R Foundation for Statistical Computing: Vienna, Austria. Available online: https://www.R-project.org/ (accessed on 2 March 2022).

24. Thomassen, S.; Riisgård, H.U. Growth and energetics of the sponge *Halichondria panicea*. *Mar. Ecol. Prog. Ser.* **1995**, *128*, 239–246. [CrossRef]

25. Larsen, P.S.; Riisgård, H.U. Pumping rate and size of demosponges—Towards an understanding using modeling. *J. Mar. Sci. Eng.* **2021**, *9*, 1308. [CrossRef]

26. Southwell, M.W.; Weisz, J.B.; Martens, C.S.; Lindquist, N. In situ fluxes of dissolved inorganic nitrogen from the sponge community on Conch Reef, Key Largo, Florida. *Limnol. Oceanogr.* **2008**, *53*, 986–996. [CrossRef]

27. Fiore, C.L.; Baker, D.M.; Lesser, M.P. Nitrogen biogeochemistry in the Carribbean sponge *Xestospongia muta*: A source or sink of dissolved inorganic nitrogen? *PLoS ONE* **2013**, *8*, e72961. [CrossRef]

28. McMurray, S.; Blum, J.E.; Pawlik, J.R. Redwood of the reef: Growth and age of the giant barrel sponge *Xestospongia muta* in the Florida Keys. *Mar. Biol.* **2008**, *155*, 159–171. [CrossRef]

29. McMurray, S.E.; Pawlik, J.R.; Finelli, C.M. Trait-mediated ecosystem impacts: How morphology and size affect pumping rates of the Caribbean giant barrel sponge. *Aquat. Biol.* **2014**, *23*, 1–13. [CrossRef]

30. Ludeman, D.A.; Reidenbach, M.A.; Leys, S.P. The energetic cost of filtration by demosponges and their behavioural response to ambient currents. *J. Exp. Biol.* **2017**, *220*, 995–1007. [CrossRef]

31. Goldstein, J.; Riisgård, H.U.; Larsen, P.S. Exhalant jet speed of single-osculum explants of the demosponge *Halichondria panicea* and basic properties of the sponge-pump. *J. Exp. Mar. Biol. Ecol.* **2019**, *511*, 82–90. [CrossRef]

32. Dahihande, A.S.; Thakur, N.L. Temperature- and size-associated differences in the skeletal structures and osculum cross-sectional area influence the pumping rate of contractile sponge *Cinachyrella* cf. cavernosa. *Mar. Ecol.* **2019**, *40*, e12565. [CrossRef]

33. Mills, D.B.; Ward, L.M.; Jones, C.; Sweeten, B.; Forth, M.; Treusch, A.H.; Canfield, D.E. Oxygen requirements of the earliest animals. *Proc. Natl. Acad. Sci. USA* **2014**, *111*, 4168–4172. [CrossRef] [PubMed]

34. Stuart, V.; Klumpp, D. Evidence for food-resource partitioning by kelp-bed filter feeders. *Mar. Ecol. Prog. Ser.* **1984**, *16*, 27–37. [CrossRef]

35. Barthel, D. On the ecophysiology of the sponge *Halichondria panicea* in Kiel Bight. I. Substrate specificity, growth and reproduction. *Mar. Ecol. Prog. Ser.* **1986**, *32*, 291–298. [CrossRef]

36. Barthel, D. On the ecophysiology of the sponge *Halichondria panicea* in Kiel Bight. II. Biomass, production, energy budget and integration in environmental processes. *Mar. Ecol. Prog. Ser.* **1988**, *43*, 87–93. [CrossRef]

37. Schiefenhövel, K.; Kunzmann, A. Sponge farming trials: Survival, attachment, and growth of two Indo-Pacific sponges, *Neopetrosia* sp. and *Stylissa massa*. *J. Mar. Biol.* **2012**, *2012*, 41736. [CrossRef]

38. Koopmans, M.; Wijffels, R.H. Seasonal Growth Rate of the Sponge *Haliclona oculata* (Demospongiae: Haplosclerida). *Mar. Biotechnol.* **2008**, *10*, 502–510. [CrossRef]

39. Frost, T.M.; Williamson, C.E. Determination of the effect of symbiotic algae on the growth of the freshwater sponge *Spongilla lacustris*. *Ecology* **1980**, *61*, 1361–1370. [CrossRef]

40. Riisgård, H.U.; Larsen, P.S. Comparative ecophysiology of active zoobenthic filter feeding, essence of current knowledge. *J. Sea Res.* **2000**, *44*, 169–193. [CrossRef]

41. Riisgård, H.U.; Larsen, P.S. Physiologically regulated valveclosure makes mussels long-term starvation survivors: Test of hypothesis. *J. Molluscan Stud.* **2015**, *81*, 303–307. [CrossRef]

42. Tang, B.; Riisgård, H.U. Physiological Regulation of Valve-Opening Degree Enables Mussels Mytilus edulis to Overcome Starvation Periods by Reducing the Oxygen Uptake. *Open J. Mar. Sci.* **2016**, *06*, 341–352. [CrossRef]

43. Kealy, R.A.; Busk, T.; Goldstein, J.; Larsen, P.S.; Riisgård, H.U. Hydrodynamic characteristics of aquiferous modules in the demosponge *Halichondria panicea*. *Mar. Biol. Res.* **2019**, *15*, 531–540. [CrossRef]

44. Goldstein, J.; Bisbo, N.; Funch, P.; Riisgård, H.U. Contraction-expansion and morphological changes of the aquiferous system in the demosponge *Halichondria panicea*. *Front. Mar. Sci.* **2020**, *7*, 113. [CrossRef]

45. Kumala, L.; Larsen, M.; Glud, R.N.; Canfield, D.E. Spatial and temporal anoxia in single-osculum *Halichondria panicea* demosponge explants studied with planar optodes. *Mar. Biol.* **2021**, *168*, 1–13. [CrossRef]

46. Riisgård, H.U. No foundation of a "3/4 power scaling law" for respiration in Biology. *Ecol. Lett.* **1998**, *1*, 71–73. [CrossRef]

47. Jørgensen, C. Fluid mechanical aspects of suspension feeding. *Mar. Ecol. Prog. Ser.* **1983**, *11*, 89–103. [CrossRef]

48. Hamburger, K.; Møhlenberg, F.; Randløv, A.; Riisgård, H.U. Size, oxygen consumption and growth in the mussel *Mytilus edulis*. *Mar. Biol.* **1983**, *75*, 303–306. [CrossRef]

49. Petersen, J.; Riisgard, H. Filtration capacity of the ascidian Ciona intestinalis and its grazing impact in a shallow fjord. *Mar. Ecol. Prog. Ser.* **1992**, *88*, 9–17. [CrossRef]

50. Frandsen, K.T.; Riisgård, H.U. Size dependent respiration and growth of jellyfish, *Aurelia aurita*. *Sarsia* **1997**, *82*, 307–312. [CrossRef]

51. Riisgard, H. Suspension feeding in the polychaete Nereis diversicolor. *Mar. Ecol. Prog. Ser.* **1991**, *70*, 29–37. [CrossRef]

52. Lüskow, F.; Kløve-Mogensen, K.; Tophøj, J.; Pedersen, L.H.; Riisgård, H.U.; Eriksen, N.T. Seasonality in lipid content of the demosponges *Halichondria panicea* and H. *bowerbanki* at two study sites in temperate Danish waters. *Front. Mar. Sci.* **2019**, *6*, 328. [CrossRef]

53. Dahihande, A.S.; Thakur, N.L. Differences in the structural components influence the pumping capacity of marine sponges. *Front. Mar. Sci.* **2021**, *8*, 671362. [CrossRef]

54. Riisgard, H.; Goldson, A. Minimal scaling of the lophophore filter-pump in ectoprocts (Bryozoa) excludes physiological regulation of filtration rate to nutritional needs. Test of hypothesis. *Mar. Ecol. Prog. Ser.* **1997**, *156*, 109–120. [CrossRef]

55. Hermansen, P.; Larsen, P.S.; Riisgård, H.U. Colony growth rate of encrusting bryozoans (*Electra pilosa* and *Celleporella hyalina*): Importance of algal concentration and water flow. *J. Exp. Mar. Biol. Ecol.* **2001**, *263*, 1–23. [CrossRef]

56. Amui-Vedel, A.-M.; Hayward, P.J.; Porter, J.S. Zooid size and growth rate of the bryozoan *Cryptosula pallasiana* Moll in relation to temperature, in culture and in its natural environment. *J. Exp. Mar. Biol. Ecol.* **2007**, *353*, 1–12. [CrossRef]

57. Key, M.M. Estimating colony age from colony size in encrusting cheilostomes. In Proceedings of the Bryozoan Studies 2019—Proceedings of the Eighteenth International Bryozoology Association Conference Liberec, Prague, Czech Republic, 16–21 June 2020.

58. Riisgård, H.U.; Larsen, P.S. Bioenergetic model and specific growth rates of jellyfish *Aurelia aurita*. *Mar. Ecol. Prog. Ser.* 2022, *in press*. [CrossRef]

Journal of
*Marine Science
and Engineering*

Review

A Review on Genus *Halichondria* (Demospongiae, Porifera)

Josephine Goldstein [1,2] and Peter Funch [2,*]

1   Marine Biological Research Centre, Department of Biology, University of Southern Denmark,
     5230 Odense M, Denmark
2   Genetics, Ecology, and Evolution, Department of Biology, Aarhus University, 8000 Aarhus C, Denmark
*   Correspondence: funch@bio.au.dk

Abstract: Demosponges of the genus *Halichondria* Fleming (1828) are common in coastal marine ecosystems worldwide and have been well-studied over the last decades. As ecologically important filter feeders, *Halichondria* species represent potentially suitable model organisms to link and fill in existing knowledge gaps in sponge biology, providing important novel insights into the physiology and evolution of the sponge holobiont. Here we review studies on the morphology, taxonomy, geographic distribution, associated fauna, life history, hydrodynamic characteristics, and coordinated behavior of *Halichondria* species.

Keywords: demosponges; morphology; taxonomy; geographic distribution; holobiont; life history; hydrodynamics; coordinated behavior; model organism

Citation: Goldstein, J.; Funch, P. A Review on Genus *Halichondria* (Demospongiae, Porifera). *J. Mar. Sci. Eng.* **2022**, *10*, 1312. https://doi.org/10.3390/jmse10091312

Academic Editor: Caterina Longo

Received: 14 August 2022
Accepted: 13 September 2022
Published: 16 September 2022

Publisher's Note: MDPI stays neutral with regard to jurisdictional claims in published maps and institutional affiliations.

1. Introduction

The genus *Halichondria* Fleming (1828) [1] (Demospongiae, Porifera; subgenera *Halichondria* and *Eumastia*) contains the most common marine sponge species of the North Atlantic [2], including the common "bread-crumb" sponge *Halichondria (Halichondria) panicea* Pallas (1766) [3] and Bowerbank's horny sponge *H. bowerbanki* Burton (1930) [4]. The most studied species, *H. panicea*, occurs in habitats covering a broad range of salinities, temperatures, turbidities, and flow conditions [5,6] and has been recorded in marine intertidal and sublittoral zones down to depths of more than 500 m [2]. *Halichondria panicea* provides substrate for many other marine organisms, including a large and varied associated fauna [7–9], symbiotic algae [10,11], and numerous bacteria [12,13]. The life histories of *Halichondria* spp. are characterized by different modes of asexual and sexual reproduction [14], with the latter revealing strong species- and habitat-specific adaptations [15–18]. *Halichondria* sponges are filter feeders capable of processing large volumes of seawater (up to six times their own body volume per minute [19]) and efficiently retaining small food particles [20], thus playing a key role in nutrient recycling of coastal marine ecosystems [8]. Modular arrangement of their leuconoid aquiferous systems [21,22] has made it possible to study the hydrodynamic properties of the sponge filter-pump, which may help to shed light on the evolution of complex filter-feeding systems in sponges (cf. [23]). Despite their apparently simple bauplan without a nervous or muscular system, *Halichondria* spp. show coordinated responses to changing environmental conditions, including phototactic responses of larvae [24], sponge body shape changes [25], and contractile behavior [22,26–28]. The detailed mechanisms underlying coordinated behavior in sponges are still unclear [29], but existing data for *Halichondria* points out the importance of cellular communication based on a neuronal-like 'toolkit' and could serve as a milestone towards an improved understanding of tissue organization in the first animals.

The vast majority of studies on *Halichondria* (a total of 11,100 research articles according to Google scholar) are based on *H. panicea* (36.4% of total research articles) with a focus on the biological and ecological aspects, whereas much fewer studies within these research fields have addressed other species, such as *H. bowerbanki* (4.0%), *H. melanadocia* Laubenfels (1936) [30] (1.5%), *H. moorei* Bergquist (1961) [31] (1.0%), or *H. semitubulosa* Lamarck (1814) [32] (0.2%, Table 1).

Table 1. Number of research articles on *Halichondria* Fleming (1828) based on genus- and species-level (cf. [33]) according to Google Scholar (Web of Science) along with the main Web of Science research categories (accessed on 6 July 2022).

Species	No. Articles	(%)	Web of Science Categories (%)
Halichondria panicea	4040 (229)	36.4	Marine Freshwater Biology (43.5), Ecology (19.2), Oceanography (16.2)
Halichondria okadai	2980 (96)	26.8	Organic Chemistry (39.6), Pharmacology Pharmacy (16.7), Biochemistry Molecular Biology (14.5)
Halichondria sp./spp.	1390 (73)	12.5	Organic Chemistry (34.3), Medicinal Chemistry (20.6), Pharmacology Pharmacy (20.6)
Genus *Halichondria*	723 (14)	6.5	Organic Chemistry (21.4), Pharmacology Pharmacy (21.4), Biochemistry Molecular Biology (14.3)
Halichondria bowerbanki	447 (10)	4.0	Ecology (50.0), Marine Freshwater Biology (40.0), Zoology (30.0)
Halichondria japonica	260 (20)	2.3	Biochemistry Molecular Biology (30.0), Organic Chemistry (20.0), Fisheries (15.0)
Halichondria cylindrata	173 (10)	1.6	Organic Chemistry (70.0), Medicinal Chemistry (30.0), Biochemistry Molecular Biology (10.0)
Halichondria melanadocia	169 (17)	1.5	Marine Freshwater Biology (52.9), Ecology (29.4), Anatomy Morphology (50.0)
Halichondria moorei	108 (2)	1.0	Marine and Freshwater Biology (50.0), Multidisciplinary Sciences (50.0)
Halichondria sitiens	89 (5)	0.8	Biodiversity Conservation (20.0), Biology (20.0), Ecology (20.0)
Halichondria oshoro	82 (2)	0.7	Microbiology (100.0)
Halichondria magniconulosa	67 (2)	0.6	Applied Chemistry (50.0), Medicinal Chemistry (50.0), Ecology (50.0)
Halichondria semitubulosa	25 (1)	0.2	Zoology (100.0)
Halichondria cartilaginea	19 (0)	0.2	-
Halichondria genitrix	19 (0)	0.2	-
Halichondria albescens	18 (0)	0.2	-
Halichondria lutea	18 (3)	0.2	Biochemistry Molecular Biology (66.7), Ecology (66.7), Evolutionary Biology (66.7)
Halichondria coerulea	14 (1)	0.1	Ecology (100.0), Marine Freshwater Biology (100.0), Oceanography (100.0)
Halichondria glabrata	14 (2)	0.1	Anatomy and Morphology (50.0), Biology (50.0), Food Science Technology (50.0)
Halichondria diazae	13 (0)	0.1	-
Halichondria cebimarensis	12 (1)	0.1	Ecology (100.0), Marine Freshwater Biology (100.0)
Halichondria phakellioides	12 (1)	0.1	Fisheries (100.0), Limnology (100.0), Marine Freshwater Biology (100.0)
Halichondria attenuata	11 (2)	0.1	Marine Freshwater Biology (50.0), Zoology (50.0)
Halichondria contorta	10 (1)	0.1	Zoology (100.0)
Halichondria topsenti	10 (0)	0.1	-
Halichondria oblonga	9 (0)	0.1	-
Halichondria aspera	8 (0)	0.1	-
Halichondria cristata	7 (0)	0.1	-
Halichondria agglomerans	5 (0)	0.0	-
Halichondria flava	5 (0)	0.0	-
Halichondria kelleri	5 (0)	0.0	-
Halichondria migottea	5 (0)	0.0	-
Halichondria osculum	5 (1)	0.0	Medicinal Chemistry (100.0), Pharmacology Pharmacy (100.0)
Halichondria colossea	4 (0)	0.0	-
Halichondria marianae	4 (2)	0.0	Marine Freshwater Biology (50.0), Zoology (50.0)
Halichondria prostrata	4 (0)	0.0	-
Halichondria tenebrica	4 (0)	0.0	-
Halichondria capensis	3 (0)	0.0	-
Halichondria convolvens	3 (0)	0.0	-
Halichondria elenae	3 (1)	0.0	Ecology (100.0), Marine Freshwater Biology (100.0)
Other species	316 (36)	2.1	Cell biology (100.0), Zoology (100.0)
Total	**11,100 (532)**	**100.0**	**Marine and Freshwater Biology (27.3), Organic Chemistry (17.5), Ecology (12.8)**

Other studies have explored the metabolite chemistry of *Halichondria*, mainly for the species *H. okadai* Kadota (1922) [34] (26.8%, Table 1), for undefined species (*Halichondria* sp./spp., 12.5%), or on a genus-level (6.5%), reflecting partially unresolved and still ongoing taxonomic revisions of *Halichondria* species [35]. Molecular biology, including studies on the sponge microbiome, has mainly been investigated on *H. okadai*, *H. japonica* Kadota (1922) [34] (2.3%), *H. cylindrata* Tanita & Hoshino (1989) [36] (1.6%), and *H. oshoro* Tanita (1961) [37] (0.7%). Few morphological studies exist for *H. melanadocia* and *H. glabrata* Keller (1891) [38] (0.1%), while research on the hydrodynamics of sponges has remained restricted to *H. panicea* and *H. coerulea* Berquist (1967) [39] (0.1%). Despite the relevance of comparative studies on sponge cell biology, most *Halichondria* species have remained understudied (2.1%, Table 1). The aim here is to provide a compilation of studies concerning sponges in the genus *Halichondria* and point out existing knowledge gaps that may aid in future studies of these ecologically important demosponges.

2. Morphology, Taxonomy, and Distribution

The genus *Halichondria* is placed in the animal phylum Porifera, class Demospongiae, subclass Heteroscleromorpha, order Suberitida, and family Halichondriidae. Growth forms of *Halichondria* species include encrusting, massive, occasionally irregularly branching, or digitate sponges with smooth or papillate surfaces. An important morphological character to separate the two subgenera, *Halichondria* and *Eumastia*, is the absence or presence of short conical papillae on the sponge surface, respectively [2]. Members of the genus *Halichondria* typically form chimneys of variable size (up to 5 cm high) with conspicuous, relatively large oscula (2–4 mm in diameter). They are characterized by their firm but compressible texture and variable color, from olive-green (due to symbiotic algae) over orange-yellow to creamy-yellow [2] (cf. Appendix A, Figure A1). The siliceous spicule skeleton of *Halichondria* consists exclusively of oxeas or oxea derivates in a wide size range, which are arranged in an ectosomal crust (200–300 μm thick) and appear scattered or in tight bundles in the choanosome along with spongin fibers [2,40]. While the functional cell morphology and number of cell types in *Halichondria* has remained largely unknown, 18 distinct cell types which comprise four major cell families, including contractile, digestive, and amoeboid-neuroid cells, have recently been described in the freshwater demosponge *Spongilla lacustris* [41].

Species identification is traditionally based on morphological characteristics, such as the shape and structure of the skeleton and the size and form of spicules [42], but several of these characters show strong intra-specific variation and are, therefore, of rather poor quality to distinguish species. For instance, a variety of growth forms are represented by *H. panicea*, ranging from thin encrusting (Figure A1a) to erect ramose (Figure A1b), which seems to depend on the intensity of ambient water currents [43] (cf. [44]). Moreover, an extensive overlap of spicule sizes in different species has been documented [2]. Molecular data used in phylogenetic studies includes complete mitochondrial genomes of several *Halichondria* species [45–47] and mitochondrial and ribosomal markers [48,49]. The classification of genus *Halichondria*, as defined in [2], is still in need of a major revision at an ordinal level [35,50], as classification based on morphology disagrees with phylogenetic analyses using molecular data. Overall, morphological, biochemical, and molecular characters applied in recent phylogenetic analyses seem to point out that *Halichondria* is nonmonophyletic [51–54].

To date, about 100 *Halichondria* species are accepted [33,55,56]. They occur in different types of marine habitats around the world, being widespread in European [4,11,57,58], American [2], and Brazilian coastal waters of the Atlantic [59], but also in parts of the Baltic Sea [60], the White Sea [61], and the Mediterranean Sea [62]. *Halichondria* species also occur in the North Pacific, including Alaska [63,64], Japan [65], Korea [42,66], and the South China Sea [67]. The closely related species *H. panicea*, *H. bowerbanki*, and other species in this complex may serve as a suitable model to illuminate possible speciation events due to their overlapping distribution in the North Atlantic, where *H. panicea* is mainly found in shallow, protected coastal regions of the eastern parts, and shows adaptation to frequent air exposure, while *H. bowerbanki* is most common in exposed habitats of the western parts, where it tolerates high levels of siltation [11]. A molecular study based on a part of the mitochondrial marker COI suggests that North East Pacific *H.* cf. *panicea* is genetically distant from and forms a sister group to a species complex consisting of European *H. panicea* and *H. bowerbanki* [53]. *Halichondria panicea* has also been reported from the Tropical Southwestern Atlantic, along with other species such as *H. magniconulosa* Hechtel (1965) [68], *H. cebimarensis*, *H. tenebrica*, *H. migottea*, *H. sulfurea* Carvalho & Hajdu (2001) [59] and *H. marianae* Santos et al. (2018) [69]. Common species in the Pacific Ocean are *H. japonica* [65], *H. okadai*, *H. oshoro* [70], *H. gageoenesis* and *H. muanensis* Kang & Sim (2008) [42], while *H. panicea* and *H. bowerbanki* have been reported from Alaska [63,64] and Korea [66], respectively. Revisions of the classification system should include more molecular data and more species and be used to reevaluate the morphological characters used in the traditional classification [50] (cf. [53,54]).

3. The Holobiont *Halichondria*

Halichondria spp. occur on a variety of inorganic and organic hard substrates, including mussel banks, small stones and rocks, and macroalgae [8,9,43,71]. The sponges themselves provide habitat for a diverse associated fauna and various symbiotic microorganisms. The associated epi- and endofauna of *H. panicea* include various Arthropoda such as skeleton shrimps (*Caprella* spp.) and copepods, but also molluscs, e.g., the scallop *Chlamys varia*, annelids, platyhelminths, and demersal fish that prey almost exclusively upon sponge epifauna [7–10]. Symbiosis with the dinoflagellate *Prorocentrum lima* has been documented in *H. okadai* [72,73], and *H. panicea* seems to harbor (intracellular) green algae [10,11]. However, many *Halichondria* species have not been investigated, indicating numerous other yet undiscovered symbiotic interactions, e.g., with dinoflagellates, cryptophytes, microalgae, and diatoms [73]. While the growth of pathogenic bacteria on *H. panicea* can cause sponge mortality under stagnant flow conditions [74], sponges harbor diverse microbial assemblages that contribute positively to host metabolism and defense [12,75,76]. *Halichondria* spp. are characterized as low microbial abundance (LMA) sponges with high variability in their bacterial diversity across species and environments [12,13,76]. While only 7 operational taxonomic units (OTUs) of microorganisms have been identified in *H. okadai* from Korea [77], about 500 OTUs were detected in *H. panicea* and *H. (Eumastia) sitiens* Schmidt (1870) [78] from the White Sea [76], respectively, and 1779 OTUs seem to be unique to *H. bowerbanki* from the mid-Atlantic region of the eastern United States [13]. The microbiome of *H. panicea* is dominated by a core taxon of Alphaproteobacteria within the class *Amylibacter* which has recently been named 'Candidatus Halichondribacter symbioticus' [12,76,79–82]. Transmission of bacterial symbionts occurs in a mixed vertical (i.e., direct through reproduction) and horizontal mode (i.e., indirect through the environment) in *H. bowerbanki*; it is likely to vary across *Halichondria* species [13]. Metagenomics have revealed that distinct viromes with low similarity to known viral sequences are associated with *H. panicea* and *H. sitiens*, suggesting the existence of bacterial antiphage systems in sponges [76].

Halichondria sponges and their microbial symbionts produce a broad spectrum of mainly symbiont-derived bioactive metabolites [83] with cytotoxic or cell growth-inhibiting properties. Substances isolated from *Halichondria* sponges include halichondrin B and okadaic acid in *H. okadai* [72,84,85] or gymnostatins and dankastatins from an *H. japonica*-derived fungal strain [86] which may additionally serve *Halichondria* sponges as a defense mechanism against pathogens, predators, and biofouling [73,87]. Okadaic acid is a biotoxin known for its cyto-, neuro-, immune-, embryo-, and genotoxicity in marine animals [87–89] and has been suggested to protect the demosponge *Suberites domuncula* from bacterial and parasitic infections [87]. Epibiotic *H. panicea* can negatively affect the heart performance of blue mussels (*Mytilus edulis*), which may be due to the sponges' release of excretory/secretory products. Such substances with cytotoxic properties and antimicrobial activity seem to benefit *H. panicea* in the competition for space and food across benthic fouling communities [90]. Neuroactive bacteria-derived compounds in *H. panicea* [73] suggest the relevance of symbiotic interactions for essential physiological processes such as coordinated behavior. The natural variability of sponge-microbe associations in *Halichondria* seems to provide a meaningful framework for modeling symbiotic interactions in metazoans (cf. [91]). In *H. bowerbanki*, for instance, changes in microbial communities after exposure to thermal stress have been documented [92], pointing out the relevance of future studies on sponges for assessing possible shifts in symbiont community composition and structure in response to global warming.

4. Life History

The life histories of *Halichondria* species typically include a reproductive period of 2–3 months in temperate regions [15,71,93]. *Halichondria* spp. are ovoviviparous and characterized by asynchronous gameto- and embryogenesis, while habitat-specific differences include successive hermaphroditism in White Sea populations of *H. panicea* and *H. sitiens* [18], simultaneous hermaphroditism in *H. panicea* and *H. bowerbanki* from the southwest coast of the Netherlands [16], incomplete gonochorism in *Halichondria* sp. from Mystic Estuary, US [15], or gonochorism in *H. panicea* from Kiel Bight, Germany [17]. In temperate regions, environmental parameters such as temperature and salinity drive the onset of sexual reproduction in *H. panicea* [17]. Differentiation of gametes from somatic cells has been observed in both *H. panicea* and *H. semitubulosa*, indicating the development of spermatocytes from choanocytes or archaeocytes, a process that may be species-dependent [62,94]. The larvae of *Halichondria* species are typically of parenchymella type and sometimes contain choanocyte chambers before settlement [24,95]. The release of *Halichondria* larvae seems to follow a light cue, being triggered by the onset of darkness in the temperate species *H. panicea* [96], while tropical *H. melanadocia* release larvae on exposure to light following a period of dark adaptation [24]. Phototactic responses of larvae range from positive to neutral to negative before settlement upon various hard substrates [24] (Figure A2a,b).

The growth of *Halichondria* sponges is dependent on temperature [70] and the concentration of available food, which mainly consists of bacteria and phytoplankton [97]. Pumping rates of *H. panicea* increase linearly with temperature and require relatively low energy demands for filtering large volumes of seawater [20,98], as expressed by F/R-ratios $\geq$15.6 L H_2O $(mL\ O_2)^{-1}$, which are comparable to other filter-feeding marine invertebrates [19]. In contrast, the energetic cost of growth is high in sponges [20,99], with exponential growth at a maximum rate of 4% d^{-1} in *H. panicea* under natural conditions [100]. The weight-specific growth of *H. panicea* is constant over sponge size, which has been pointed out as a unique feature among most other filter-feeding invertebrates, reflecting the modular organization of sponges [100]. A study of *H. panicea* from the Western Baltic Sea suggested that stored glycogen reserves fueled sexual reproduction and that the sponges degenerated in the end of the following year after reproduction [71]. Tissue regression and high mortality during the colder months of the year have also been reported for temperate *Halichondria* sp. from the Mystic and Thames estuaries, US [57,101] and for *Halichondria bowerbanki* from New England, US [102], respectively, while the longevity of *H. okadai* in Japanese waters may exceed 3 years when considering asexual reproduction, i.e., fission and fusion of sponge fragments [14]. *Halichondria panicea* is capable of rapid regeneration of damaged parts, as expressed in $\geq$3-fold increased growth rates in response to predation [103] or during the reorganization of the aquiferous system in explant cuttings within approximately 6–10 days [22] (Figs. A2c-f), while other species, such as *H. magniconulosa*, seem to regenerate at slower rates [104]. Several *Halichondria* species, including *H. lutea* Alcolado (1984) [105], *H. magniconulosa*, and *H. melanadocia* have been recognized as important members of the Caribbean mangrove and coral reef communities, where they are preyed upon by fish [106,107]. *H. panicea* can also serve as a food source for hermit crabs, shrimp, large isopods (e.g., *Idothea* sp.), or the nudibranch *Archidoris montereyensis*, which may appear in such high density that it can eliminate large and long-lived sponge populations [63,64]. *Halichondria* sponges play an important role in nutrient recycling of coastal marine ecosystems due to their unique ability to retain small particles ($\leq$0.1 μm) [20,108]. Regular tissue sloughing has been observed in *H. panicea* in response to sedimentation of organic material and settlement of small organisms on the sponge surface [109], along with seasonal remineralization of released *H. panicea* biomass following reproduction [8]. As the water pumping activity of *H. panicea* leads to an accumulation of pollutants, such as heavy metals, in direct proportion to ambient concentrations, their potential use as biomonitoring organisms has been proposed [40,110].

5. Hydrodynamics

As for other demosponges, the aquiferous system of *Halichondria* is leuconoid [21,40,111] and characterized by choanocytes organized in small spherical chambers which create a unidirectional flow of ambient water through a complex canal system [112,113]. The aquiferous elements of *Halichondria* act like a sieve for particles of variable size due to their aperture diameters (Figure A3a). As documented for *H. panicea*, they include numerous inhalant openings (ostia; 7–32 μm) through which seawater is drawn into incurrent canals (50–200 μm), finer incurrent canal branches (prosodi; 5 μm), and the prosopyles (1–4 μm) of choanocyte chambers (18–35 μm; Figure A3b) [113]. Here, choanocytes retain small food particles $\leq$0.1 μm [20] on their microvilli collars (Figure A4a). Each choanocyte chamber of *H. panicea* contains about 40–120 choanocytes at an estimated choanocyte chamber density of 18,000 mm^{-3} [113]. Water leaves choanocyte chambers through an apopyle (7–17 μm; Figure A4b) via excurrent canals (140–450 μm), which drain into an atrium (2.1 mm) from where the water exits the sponge in an excurrent jet through the osculum (1.2 mm) [113] (but see also [21]).

Each osculum represents a functional unit of aquiferous elements in a certain sponge volume (cf. Figure A2b–d), thus characterizing *Halichondria* sponges with multiple oscula as an array of several autonomous aquiferous modules [22,114,115]. The pumping rate of each aquiferous module is directly proportional to the density of choanocyte chambers in *H. panicea* [22], implying constant choanocyte densities for different *Halichondria* species. However, module size seems to determine the volume-specific pumping rates of *H. panicea*, which can reach a maximum of 15 mL min^{-1} (cm^3 sponge)$^{-1}$ in growing modules, as observed in single-osculum explants [26,27] (Figure A2c,d), while the modules in multi-oscula explants seem to pump at a lower maximum rate of 3 mL min^{-1} (cm^3 sponge)$^{-1}$ [22], probably due to a decrease in choanocyte chamber density with increasing module volume [116]. Based on the volume-specific pumping rate and choanocyte chamber density of *H. panicea*, the pumping rate per choanocyte chamber in a multi-oscula sponge can be estimated to $(3/18,000)/60 = 2.78 \times 10^{-6}$ mm^3 s^{-1} = 2778 μm^{-3} s^{-1}, and thus the pumping rate per choanocyte at an average of 80 choanocytes per chamber [113], to (2778/80 = 35 μm^3 s^{-1}). This value is in range with a previous estimate of $(4.46 \times 10^{-6}$ mm^3 s$^{-1}/95 = 47$ μm^3 s$^{-1})$ for the demosponge *Haliclona permollis* [113,117] (their Table 1, respectively). A recent hydrodynamic model on the pump characteristics of leuconoid sponges assumed the presence of flagellar vanes along with a glycocalyx mesh which distally connects the microvilli collars of choanocytes, as has been shown for the freshwater sponge *Spongilla lacustris* [118,119], in order to deliver observed pump pressures [23]. These ultra-structural features of choanocytes have so far not been documented in *Halichondria* (cf. Figure A4a), pointing out the need for further studies on ultrastructure and hydrodynamic properties, which may provide valuable insight into the evolution of demosponge filter-pump systems (cf. [120]).

6. Coordinated Behavior

At least three different basic cell types are found in *Halichondria* species, including choanocytes, pinacocytes, and amoeboid (mesohyl) cells [24,121]. The coordinated behavior of sponge cells mediates the hydrodynamic and physical properties of the aquiferous system required for efficient filter feeding under different environmental conditions. Communication between motile cells is the basic principle underlying continuous tissue reorganization, regeneration, and microscale movements in sponges [122–125]—a topic which has, unfortunately, so far only been addressed by a few studies on *Halichondria* spp. Continuous tissue remodeling in *H. panicea*, as expressed by fusion, shape changes, and movement of sponges, has been observed in aquaria and intertidal rocky pools [25]. *Halichondria japonica* explants have been shown to fuse with explants of the same sponge, while they reject cells from other *H. japonica* sponges or from *H. okadai* [126]. Several types of mesohyl cells seem to be involved in this process of "self and nonself" recognition in *H. japonica*, including amoeboid archaeocytes, motile (granule-rich) gray cells, and collencytes [126].

Recent work on *H. panicea* points out the importance of cellular transport for the removal of inedible particles from the aquiferous system [27]. At the same time, sponge sandwich cultures may provide a suitable method (Figure A2e,f) for studying the cell types and mechanisms mediating capture, transport, and digestion/removal of edible and inedible particles (Figure A5).

Coordinated behavior further includes contraction of various parts of the aquiferous system, including the osculum [26], in- and excurrent canals, ostia and apopylar openings of the choanocyte chambers, resulting in reduction and temporal shut down of the water flow through single-osculum explants of *H. panicea* [27,28]. Contractile behavior is common among sponges and seems to follow species-specific cycles of distinct frequency and intensity [127–131] which can be expressed in asynchronous patterns across conspecifics in *H. panicea* [28,132]. Contractions can occur spontaneously in *H. panicea* explants under undisturbed conditions in the laboratory and can be induced by chemical messengers (γ-aminobutyric acid and L-glutamate) or by mechanical stimulation with inedible particles [28]. Coordinated contractions of different aquiferous modules in *H. panicea* explants with multiple oscula have been observed in response to external stimuli [22]. Peristaltic-like waves of contraction travel through the sponge, resulting in osculum closure at speeds of up to 233 nm s^{-1} in *H. panicea* (15 °C) [28]. Comparatively, observed contraction speeds of up to 12 µm s^{-1} in the marine demosponge *Tethya wilhema* (26 °C) [129] and 122 µm s^{-1} in the freshwater demosponge *Ephydatia muelleri* (21 °C) [131] seem considerably higher, emphasizing the need for future studies on the contraction kinetics of *Halichondria* species. During contractions, *H. panicea* shows reduced pumping activity [19,26,27], an associated decrease in respiration rates [132], and local internal oxygen depletion [133]. These physiological changes have been suggested as adaptations to variable environmental conditions, including food limitation [134], resuspension of sediment during storm events [135] (cf. [136]), seasonal changes in water temperature, changes in illumination period, spawning events of other sponge species [128], and facilitation of suitable habitat for specific symbiotic microorganisms [132,133]. Contractions may serve *Halichondria* sponges as an important mechanism to protect the sponge filter-pump in distinct aquiferous modules from clogging and damage and seem to be mediated by exo- and endopinacocytes [22,27,28,134,137], while the underlying cellular pathways have remained unclear. Previous studies have described contractile epithelial cells in sponges that function based on a 'toolkit' of neuronal-like elements, including sensory cilia, conduction pathways, and signaling molecules [41,134,138–140]. The pinacocytes of other demosponges exhibit actomyosin-based contractility [41,130,137,139,141,142], and myosin type II has been isolated from cells of *H. okadai* [143].

It is likely that communication between sponge cells in *Halichondria* is based on the extracellular spreading of chemical messengers [41,123,144], neuronal-like receptors [145], and cell contacts via cellular processes/membrane junctions [146–148]. As the abovementioned examples emphasize, cellular communication pathways require further attention in future studies. More detailed information on the functional cell morphology of *Halichondria*, as can now be accessed using whole-body single-cell RNA sequencing (cf. [41]), is needed to shed light on the principles underlying coordinated behavior in sponges. We encourage future work on the LMA demosponge *H. panicea* as a model organism to revisit functional coordination pathways with an integral perspective on the underlying morphological structures combining molecular, cytological, and physiological techniques.

7. Conclusions

Halichondria sponges are well-studied and the literature represents a strong base for our present understanding of the ecology and physiology of demosponges. Previous work has mainly focused on *H. panicea*, paving the foundations for modeling sponge-microbe interactions, hydrodynamics of the sponge filter pump, and cell communication in demosponges. We encourage future research to fill in present knowledge gaps regarding the functional cell morphology and filter-pump characteristics of *H. panicea*, along with comparative studies including other *Halichondria* species, to improve and verify existing models based on this ubiquitous demosponge genus.

Author Contributions: Conceptualization, J.G. and P.F.; writing—original draft preparation, J.G.; writing—review & editing, P.F.; visualization, J.G.; project administration, P.F.; funding acquisition, P.F. All authors have read and agreed to the published version of the manuscript.

Funding: This research was funded by the Independent Research Fund, grant number 8021-00392B.

Institutional Review Board Statement: Not applicable.

Informed Consent Statement: Not applicable.

Data Availability Statement: Not applicable.

Acknowledgments: We are grateful to three anonymous reviewers for providing valuable feedback on the manuscript. We further thank Héloïse Hamel and Janni Magelund Degn Larsen for supplementary photographs. Stereo-, light and scanning electron microscopy (SEM) images were acquired at the Marine Biological Research Centre, Kerteminde, University of Southern Denmark, and at the Interdisciplinary Nanoscience Center (iNANO), Aarhus University, Denmark.

Conflicts of Interest: The authors declare no conflict of interest.

Appendix A

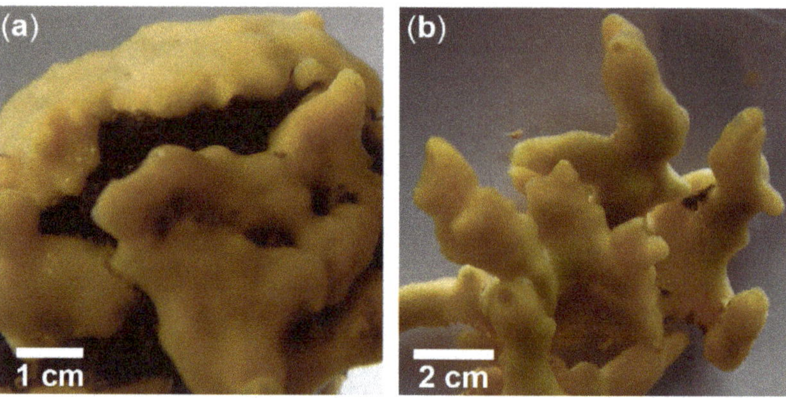

Figure A1. Growth forms of *Halichondria panicea* Pallas (1766) in the inlet to Kerteminde Fjord, Denmark (55°26′59″ N, 10°39′41″ E). (**a**) Growing on a piece of rope, collected in November 2020 and (**b**) with finger-shaped chimneys, found on wood in November 2020. Pictures: Héloïse Hamel.

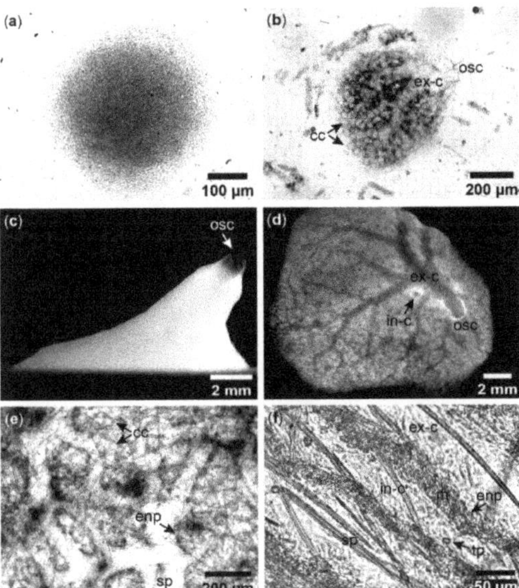

Figure A2. Aquiferous module formation in *Halichondria panicea*. (**a**) Sponge cells after larval settlement, (**b**) development of choanocyte chambers (cc), excurrent canals (ex-c) and an osculum (osc) in a juvenile sponge, (**c**) single-osculum explant (side-view), (**d**) explant (top-view) with visible incurrent (in-c) and excurrent canals (ex-c), (**e**) sandwich culture with choanocyte chambers (cc), spicules (sp), and endopinacoderm (enp) lining aquiferous canals, (**f**) sandwich culture after addition of edible particles (tp) for tracing water flow in the incurrent canal (in-c) which is separated from the flow in the excurrent canal (ex-c) by endopinacocytes (enp) and mesohyl (m).

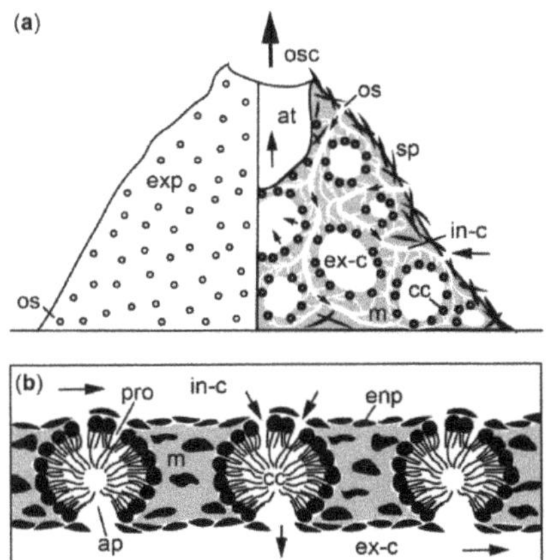

Figure A3. Schematic illustration of the aquiferous system in a functional module of *Halichondria panicea*. (**a**) Left: external surface with ostia (open circles), right: canal system with choanocyte chambers (black circles) and flow direction towards osculum indicated by arrows (**b**) water flow (arrows) through choanocyte chambers (cf. [111,117], their Figures 9d and 2b, respectively). Abbreviations: exp = exopinacoderm, os = ostium, in-c = incurrent canal, enp = endopinacoderm, pro = prosopyle, cc = choanocyte chamber, ap = apopyle, m = mesohyl, sp = spicule, ex-c = excurrent canal, at = atrium, osc = osculum.

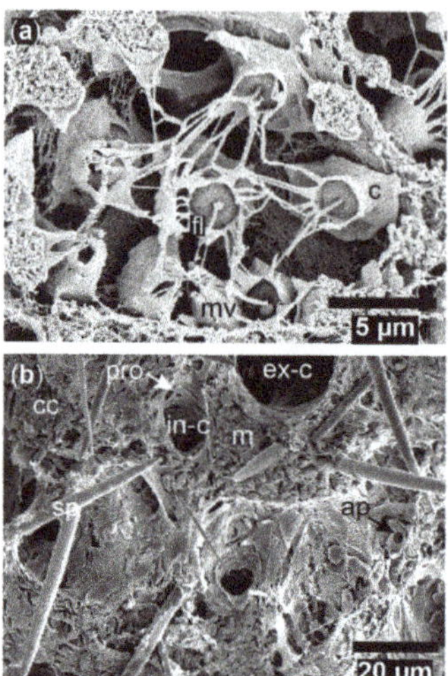

Figure A4. *Halichondria panicea.* SEM of cryo-fractured explants. (**a**) Choanocyte chamber with choanocytes (c) and their microvilli collars (mv) surrounding the flagellum (fl), (**b**) the fracture shows components of the aquiferous system with prosopyles (pro) and apopyles (ap) connected to incurrent (in-c) and excurrent canals (ex-c), respectively, embedded in mesohyl (m) with choanocyte chambers (cc) and spicules (sp).

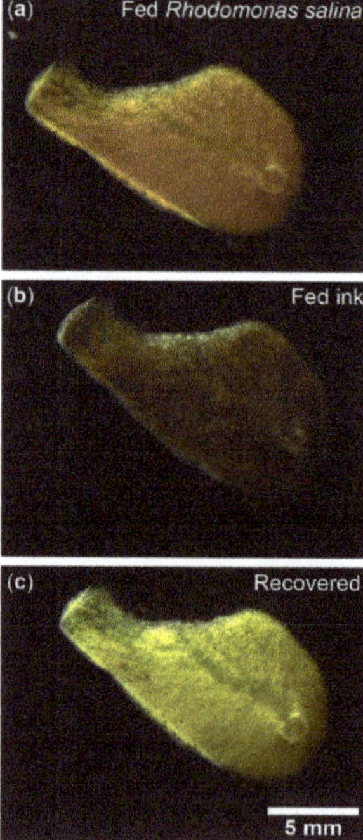

Figure A5. Exposure of *Halichondria panicea* to different particle types. Single-osculum explant (top-view) after (**a**) feeding on *Rhodomonas salina* (Cryptophyceae); note the red color originating from added algae, (**b**) exposure to inedible ink (Pelikan Scribtol, 2×10^4-fold diluted) for 1 h; note black color, and (**c**) recovery in particle-free seawater for 24 h. Pictures: Janni Magelund Degn Larsen.

References

1. Fleming, J. *A History of British Animals: Exhibiting the Descriptive Characters and Systematical Arrangement of the Genera and Species of Quadrupeds, Birds, Reptiles, Fishes, Mollusca, and Radiata of the United Kingdom, Including the Indigenous, Extirpated, and Extinct Kinds, Together with Periodical and Occasional Visitors*; Bell & Bradfute: London, UK, 1828; pp. 506–524.
2. Erpenbeck, D.; Van Soest, R.W. Family Halichondriidae Gray, 1867. In *Systema Porifera*, 1st ed.; Hooper, J.N.A., Van Soest, R.W.M., Willenz, P., Eds.; Springer: Boston, MA, USA, 2002; pp. 787–815.
3. Pallas, P.S. *Elenchus Zoophytorum Sistens Generum Adumbrationes Generaliores et Specierum Cognitarum Succintas Descriptiones, Cum Selectis Auctorum Synonymis*; Fransiscum Varrentrapp: The Hague, The Netherlands, 1766; p. 451.
4. Burton, M. Additions to the sponge fauna at Plymouth. *J. Mar. Biolog. Assoc. U. K.* **1930**, *16*, 489–508. [CrossRef]
5. Hiscock, K.; Jones, H. Halichondria (Halichondria) bowerbanki Bowerbank's halichondria. In *Marine Life Information Network: Biology and Sensitivity Key Information Reviews*; Tyler-Walters, H., Hiscock, K., Eds.; Marine Biological Association of the United Kingdom: Plymouth, UK, 2007; pp. 1–14. [CrossRef]
6. Hiscock, K. Halichondria (Halichondria) panicea breadcrumb sponge. In *Marine Life Information Network: Biology and Sensitivity Key Information Reviews*; Tyler-Walters, H., Hiscock, K., Eds.; Marine Biological Association of the United Kingdom: Plymouth, UK, 2008; pp. 1–16. [CrossRef]
7. Forester, A.J. The association between the sponge *Halichondria* panicea (Pallas) and scallop *Chlamys varia* (L.): A commensal-protective mutualism. *J. Exp. Mar. Biol. Ecol.* **1979**, *36*, 1–10. [CrossRef]
8. Barthel, D. On the ecophysiology of the sponge *Halichondria* panicea in Kiel Bight. 2. Biomass, production, energy budget and integration in environmental processes. *Mar. Ecol. Prog. Ser.* **1988**, *43*, 87–93. [CrossRef]
9. Peattie, M.E.; Hoare, R. The sublittoral ecology of the Menai Strait: II. The sponge *Halichondria* panicea (Pallas) and its associated fauna. *Estuar. Coast. Shelf Sci.* **1981**, *13*, 621–635. [CrossRef]
10. Long, E.R. The associates of four species of marine sponges of Oregon and Washington. *Pac. Sci.* **1968**, *22*, 347–351.
11. Vethaak, A.D.; Cronie, R.J.A.; Van Soest, R.W.M. Ecology and distribution of two sympatric, closely related sponge species, *Halichondria* panicea (Pallas, 1766) and H. bowerbanki Burton, 1930 (Porifera, Demospongiae), with remarks on their speciation. *Bijdr. Dierkd.* **1982**, *52*, 82–102. [CrossRef]
12. Knobloch, S.; Jóhannsson, R.; Marteinsson, V. Bacterial diversity in the marine sponge *Halichondria* panicea from Icelandic waters and host-specificity of its dominant symbiont "Candidatus Halichondribacter symbioticus". *FEMS Microbiol. Ecol.* **2019**, *95*, fiy220. [CrossRef]
13. Sacristán-Soriano, O.; Winkler, M.; Erwin, P.; Weisz, J.; Harriott, O.; Heussler, G.; Hill, M. Ontogeny of symbiont community structure in two carotenoid-rich, viviparous marine sponges: Comparison of microbiomes and analysis of culturable pigmented heterotrophic bacteria. *Environ. Microbiol. Rep.* **2019**, *11*, 249–261. [CrossRef]
14. Tanaka, K. Growth dynamics and mortality of the intertidal encrusting sponge *Halichondria* okadai (Demospongiae, Halichondrida). *Mar. Biol.* **2002**, *140*, 383–389.
15. Fell, P.E.; Jacob, W.F. Reproduction and development of *Halichondria* sp. in the Mystic Estuary, Connecticut. *Biol. Bull.* **1979**, *156*, 62–75. [CrossRef]
16. Wapstra, M.; Van Soest, R.W.M. Sexual reproduction, larval morphology and behaviour in demosponges from the southwest of the Netherlands. In *Taxonomy of Porifera*, 1st ed.; Vacelet, J., Boury-Esnault, N., Eds.; Springer: Berlin/Heidelberg, Germany, 1987; pp. 281–307.
17. Witte, U.; Barthel, D.; Tendal, O. The reproductive cycle of the sponge *Halichondria* panicea Pallas (1766) and its relationship to temperature and salinity. *J. Exp. Mar. Biol. Ecol.* **1994**, *183*, 41–52. [CrossRef]
18. Gerasimova, E.I.; Ereskovsky, A.V. Reproduction of two species of *Halichondria* (Demospongiae: Halichondriidae) in the White Sea. In *Porifera Research—Biodiversity, Innovation and Sustainability. Série Livros*, 1st ed.; Custódio, M.R., Ed.; Museu Nacional: Rio de Janeiro, Brazil, 2007; Volume 28, pp. 327–333.
19. Riisgård, H.U.; Kumala, L.; Charitonidou, K. Using the F/R-ratio for an evaluation of the ability of the demosponge *Halichondria* panicea to nourish solely on phytoplankton versus free-living bacteria in the sea. *Mar. Biol. Res.* **2016**, *12*, 907–916. [CrossRef]
20. Thomassen, S.; Riisgård, H.U. Growth and energetics of the sponge *Halichondria* panicea. *Mar. Ecol. Prog. Ser.* **1995**, *128*, 239–246. [CrossRef]
21. Vogel, S. Current-induced flow through the sponge, Halichondria. *Biol. Bull.* **1974**, *147*, 443–456. [CrossRef]
22. Kealy, R.A.; Busk, T.; Goldstein, J.; Larsen, P.S.; Riisgård, H.U. Hydrodynamic characteristics of aquiferous modules in the demosponge *Halichondria* panicea. *Mar. Biol. Res.* **2019**, *15*, 531–540. [CrossRef]
23. Asadzadeh, S.S.; Larsen, P.S.; Riisgård, H.U.; Walther, J.H. Hydrodynamics of the leucon sponge pump. *J. R. Soc. Interface* **2019**, *16*, 20180630. [CrossRef]
24. Woollacott, R.M. Structure and swimming behavior of the larva of *Halichondria* melanadocia (Porifera: Demospongiae). *J. Morphol.* **1990**, *205*, 135–145. [CrossRef] [PubMed]
25. Burton, M. Observations on littoral sponges, including the supposed swarming of larvae, movement and coalescence in mature individuals, longevity and death. *Proc. Zool. Soc. Lond.* **1949**, *118*, 893–915. [CrossRef]
26. Kumala, L.; Riisgård, H.U.; Canfield, D.E. Osculum dynamics and filtration activity studied in small single-osculum explants of the demosponge *Halichondria* panicea. *Mar. Ecol. Prog. Ser.* **2017**, *572*, 117–128. [CrossRef]

27. Goldstein, J.; Riisgård, H.U.; Larsen, P.S. Exhalant jet speed of single-osculum explants of the demosponge*Halichondria*panicea and basic properties of the sponge-pump. *J. Exp. Mar. Biol. Ecol.* **2019**, *511*, 82–90. [CrossRef]

28. Goldstein, J.; Bisbo, N.; Funch, P.; Riisgård, H.U. Contraction-expansion and the effects on the aquiferous system in the demosponge*Halichondria*panicea. *Front. Mar. Sci.* **2020**, *7*, 113. [CrossRef]

29. Abe, T.; Sahin, F.P.; Akiyama, K.; Naito, T.; Kishigami, M.; Miyamoto, K.; Sakakibara, Y.; Uemura, D. Construction of a metagenomic library for the marine sponge*Halichondria*okadai. *Biosc. Biotechnol. Biochem.* **2012**, *76*, 633–639. [CrossRef]

30. Laubenfels, M.W. A discussion of the sponge fauna of the Dry Tortugas in particular and the West Indies in general, with material for a revision of the families and orders of the Porifera. *Publ. Carnegie Instit. Wash.* **1936**, *467*, 1–225.

31. Bergquist, P.R. A collection of Porifera from Northern New Zealand, with descriptions of seventeen new species. *Pac. Sci.* **1961**, *15*, 33–48.

32. Lamarck, J.B. Sur les polypiers empâtés. *Ann. Mus. Natl. Hist. Nat.* **1814**, *20*, 294–312.

33. Van Soest, R.W.; Boury-Esnault, N.; Hooper, J.N.; Rützler, K.; de Voogd, N.J.; de Glasby, B.A.; Hajdu, E.; Pisera, A.B.; Manconi, R.; Schoenberg, C.; et al. Halichondria Fleming, 1828. World Porifera Database. World Register of Marine Species 2020. Available online: https://www.marinespecies.org/aphia.php?p=taxdetails&id=131807 (accessed on 6 July 2022).

34. Kadota, J. Observations of two new species of the genus Reniera of monaxonid sponges. *Zool. Mag.* **1922**, *34*, 700–711.

35. Erpenbeck, D.; Hall, K.; Alvarez, B.; Büttner, G.; Sacher, K.; Schätzle, S.; Schuster, A.; Vargas, S.; Hooper, J.N.A.; Wörheide, G. The phylogeny of halichondrid demosponges: Past and present re-visited with DNA-barcoding data. *Org. Divers. Evol.* **2012**, *12*, 57–70. [CrossRef]

36. Tanita, S.; Hoshino, T. *The Demospongiae of Sagami Bay*; Biological Laboratory, Imperial Household: Tokyo, Japan, 1989; p. 197.

37. Tanita, S. Two sponges obtained by the Training Ship 'Oshoro Maru' in the Eastern Behring Sea. *Bull. Fish. Sci. Hokkaido Univ.* **1961**, *11*, 183–187.

38. Keller, C. Die Spongienfauna des Rothen Meeres (II. Hälfte). *Z. Wiss. Zool.* **1891**, *52*, 294–368.

39. Bergquist, P.R. Additions to the sponge fauna of the Hawaiian Islands. *Micronesica* **1967**, *3*, 159–174.

40. Olesen, T.M.E.; Weeks, J.M. Accumulation of Cd by the marine sponge*Halichondria*panicea Pallas: Effects upon filtration rate and its relevance for biomonitoring. *Bull. Environ. Contam. Toxicol.* **1994**, *52*, 722–728. [CrossRef] [PubMed]

41. Musser, J.M.; Schippers, K.J.; Nickel, M.; Mizzon, G.; Kohn, A.B.; Pape, C.; Ronchi, P.; Papadopoulos, N.; Tarashansky, A.J.; Hammel, J.U.; et al. Profiling cellular diversity in sponges informs animal cell type and nervous system evolution. *Science* **2021**, *374*, 717–723. [CrossRef]

42. Kang, D.W.; Sim, C.J. Two new sponges of the genus*Halichondria*(Halichondrida: Halichondriidae) from Korea. *Anim. Cells Syst.* **2008**, *12*, 65–68. [CrossRef]

43. Barthel, D. Influence of different current regimes on the growth form of*Halichondria*panicea Pallas. In *Fossil and Recent Sponges*, 1st ed.; Reitner, J., Keupp, H., Eds.; Springer: Berlin/Heidelberg, Germany, 1991; pp. 387–394.

44. Bell, J.J.; Barnes, D.K. The influences of bathymetry and flow regime upon the morphology of sublittoral sponge communities. *J. Mar. Biolog. Assoc. U. K.* **2000**, *80*, 707–718. [CrossRef]

45. Wang, D.; Zhang, Y.; Huang, D. The complete mitochondrial genome of sponge*Halichondria*(Halichondria) sp. (Demospongiae, Suberitida, Halichondriidae). *Mitochondrial DNA B Resour.* **2016**, *1*, 512–514. [CrossRef]

46. Kim, H.; Kim, H.J.; Jung, Y.H.; Yu, C.; An, Y.R.; Han, D.; Kang, D.W. The complete mitochondrial genome of sponge*Halichondria*okadai (Demospongiae, Suberitida, Halichondriidae) from Korea water. *Mitochondrial DNA B Resour.* **2017**, *2*, 873–874. [CrossRef]

47. Kim, H.; Kang, D.W.; Yu, C.; Jung, Y.H.; Yoon, M.; Kim, H.J. The complete mitochondrial genome of sponge*Halichondria*sp. (Demospongiae, Suberitida, Halichondriidae) from Dok-do. *Mitochondrial DNA B Resour.* **2019**, *4*, 1729–1730. [CrossRef]

48. Erpenbeck, D.; Breeuwer, J.; van der Velde, H.; van Soest, R. Unravelling host and symbiont phylogenies of halichondrid sponges (Demospongiae, Porifera) using a mitochondrial marker. *Mar. Biol.* **2002**, *141*, 377–386.

49. Erpenbeck, D.; Duran, S.; Rützler, K.; Paul, V.J.; Hooper, J.N.; Wörheide, G. Towards a DNA taxonomy of Caribbean demosponges: A gene tree reconstructed from partial mitochondrial CO1 gene sequences supports previous rDNA phylogenies and provides a new perspective on the systematics of Demospongiae. *J. Mar. Biolog. Assoc. U. K.* **2007**, *87*, 1563–1570. [CrossRef]

50. Morrow, C.C.; Picton, B.E.; Erpenbeck, D.; Boury-Esnault, N.; Maggs, C.A.; Allcock, A.L. Congruence between nuclear and mitochondrial genes in Demospongiae: A new hypothesis for relationships within the G4 clade (Porifera: Demospongiae). *Mol. Phylogenetics Evol.* **2012**, *62*, 174–190. [CrossRef]

51. Alvarez, B.; Crisp, M.D.; Driver, F.; Hooper, J.N.; Van Soest, R.W. Phylogenetic relationships of the family Axinellidae (Porifera: Demospongiae) using morphological and molecular data. *Zool. Scr.* **2000**, *29*, 169–198. [CrossRef]

52. Castellanos, L.; Zea, S.; Osorno, O.; Duque, C. Phylogenetic analysis of the order Halichondrida (Porifera, Demospongiae), using 3β-hydroxysterols as chemical characters. *Biochem. Syst. Ecol.* **2003**, *31*, 1163–1183. [CrossRef]

53. Erpenbeck, D. A molecular comparison of Alaskan and North East Atlantic*Halichondria*panicea (Pallas 1766) (Porifera: Demospongiae) populations. *Boll. Mus. Ist. Biol. Univ. Genova* **2004**, *68*, 319–325.

54. Erpenbeck, D.; Breeuwer, J.A.; Van Soest, R.W. Identification, characterization and phylogenetic signal of an elongation factor-1 alpha fragment in demosponges (Metazoa, Porifera, Demospongiae). *Zool. Scr.* **2005**, *34*, 437–445. [CrossRef]

55. Hooper, J.N.A.; van Soest, R.W.M. *Systema Porifera. A Guide to the Classification of Sponges*; Kluwer Academic/Plenum Publishers: New York, NY, USA, 2002; Volume 1, pp. 1–1101.

56. Alvarez, B.; Hooper, J.N. Taxonomic revision of the order Halichondrida (Porifera: Demospongiae) of northern Australia. Family Halichondriidae. *Beagle Rec. Mus. Art Galleries North. Territ.* **2011**, *27*, 55–84. [CrossRef]

57. Fell, P.E.; Lewandrowski, K.B. Population dynamics of the estuarine sponge,*Halichondria*sp., within a New England eelgrass community. *J. Exp. Mar. Biol. Ecol.* **1981**, *55*, 49–63. [CrossRef]

58. Picton, B.E.; Goodwin, C.E. Sponge biodiversity of Rathlin Island, Northern Ireland. *J. Mar. Biolog. Assoc. U. K.* **2007**, *87*, 1441–1458. [CrossRef]

59. Carvalho, M.D.S.; Hajdu, E. Comments on brazilian*Halichondria*Fleming (Halichondriidae, Halichondrida, Demospongiae), with the description of four new species from the São Sebastião Channel and its environs (Tropical Southwestern Atlantic). *Rev. Bras. Zool.* **2001**, *18*, 161–180. [CrossRef]

60. Barthel, D. Population dynamics of the sponge*Halichondria*panicea (Pallas) in Kiel Bight. In *Marine Eutrophication and Population Dynamics: 25th European Marine Biology Symposium*; Colombo, G., Ed.; Olsen & Olsen: Fredensborg, Denmark, 1992; pp. 203–209.

61. Khalaman, V.V.; Komendantov, A.Y. Experimental study of the ability of the sponge*Halichondria*panicea (Porifera: Demospongiae) to compete for a substrate in shallow-water fouling communities of the White Sea. *Biol. Bull. Russ. Acad. Sci.* **2016**, *43*, 69–74. [CrossRef]

62. Gaino, E.; Lepore, E.; Rebora, M.; Mercurio, M.; Sciscioli, M. Some steps of spermatogenesis in*Halichondria*semitubulosa (Demospongiae, Halichondriidae). *Ital. J. Zool.* **2007**, *74*, 117–122. [CrossRef]

63. Knowlton, A.L.; Highsmith, R.C. Convergence in the time-space continuum: A predator-prey interaction. *Mar. Ecol. Prog. Ser.* **2000**, *197*, 285–291. [CrossRef]

64. Wulff, J. Ecological interactions and the distribution, abundance, and diversity of sponges. *Adv. Mar. Biol.* **2012**, *61*, 273–344.

65. Hoshino, S. Systematic status of*Halichondria*japonica (Kadota) (Demospongiae, Halichondrida) from Japan. *Boll. Mus. Ist. Biol. Univ. Genova* **2004**, *68*, 373–379.

66. Jeon, Y.J.; Sim, C.J. A new record of genus*Halichondria*(Demospongiae: Halichondrida: Halichondriidae) from Korea. *Anim. Syst. Evol. Diversity* **2009**, *25*, 137–139. [CrossRef]

67. Li, Z.; Hu, Y.; Liu, Y.; Huang, Y.; He, L.; Miao, X. 16S rDNA clone library-based bacterial phylogenetic diversity associated with three South China Sea sponges. *World J. Microbiol. Biotechnol.* **2007**, *23*, 1265–1272. [CrossRef]

68. Hechtel, G.J. A systematic study of the Demospongiae of Port Royal, Jamaica. *Bull. Peabody Mus. Nat. Hist.* **1965**, *20*, 1–103.

69. Santos, G.G.; Nascimento, E.; Pinheiro, U. Halichondriidae Gray, 1867 from the Northeastern Brazil with description of a new species. *Zootaxa* **2018**, *4379*, 556–566. [CrossRef]

70. Kim, H.S.; Parkc, B.J.; Sim, C.J. Marine sponges in South Korea (I). *Korean J. Syst. Zool.* **1986**, *11*, 37–47.

71. Barthel, D. On the ecophysiology of the sponge*Halichondria*panicea in Kiel Bight. I. Substrate specificity, growth and reproduction. *Mar. Ecol. Prog. Ser.* **1986**, *32*, 291–298. [CrossRef]

72. Kobayashi, J.; Ishibashi, M. Bioactive metabolites of symbiotic marine microorganisms. *Chem. Rev.* **1993**, *93*, 1753–1769. [CrossRef]

73. Lee, Y.K.; Lee, J.H.; Lee, H.K. Microbial symbiosis in marine sponges. *J. Microbiol.* **2001**, *39*, 254–264.

74. Hummel, H.; Sepers, A.B.; De Wolf, L.; Melissen, F.W. Bacterial growth on the marine sponge*Halichondria*panicea induced by reduced waterflow rate. *Mar. Ecol. Prog. Ser.* **1988**, *42*, 195–198. [CrossRef]

75. Hentschel, U.; Hopke, J.; Horn, M.; Friedrich, A.B.; Wagner, M.; Hacker, J.; Moore, B.S. Molecular evidence for a uniform microbial community in sponges from different oceans. *Appl. Environ. Microbiol.* **2002**, *68*, 4431–4440. [CrossRef]

76. Rusanova, A.; Fedorchuk, V.; Toshchakov, S.; Dubiley, S.; Sutormin, D. An interplay between viruses and bacteria associated with the White Sea sponges revealed by metagenomics. *Life* **2021**, *12*, 25. [CrossRef]

77. Jeong, J.B.; Kim, K.H.; Park, J.S. Sponge-specific unknown bacterial groups detected in marine sponges collected from Korea through barcoded pyrosequencing. *J. Microbiol. Biotechnol.* **2015**, *25*, 1–10. [CrossRef]

78. Schmidt, O. *Grundzüge einer Spongien-Fauna des atlantischen Gebietes*; Wilhelm Engelmann: Leipzig, Germany, 1870; pp. 1–88.

79. Wichels, A.; Würtz, S.; Döpke, H.; Schütt, C.; Gerdts, G. Bacterial diversity in the breadcrumb sponge *Halichondria*panicea (Pallas). *FEMS Microbiol. Ecol.* **2006**, *56*, 102–118. [CrossRef]

80. Steinert, G.; Rohde, S.; Janussen, D.; Blaurock, C.; Schupp, P.J. Host-specific assembly of sponge-associated prokaryotes at high taxonomic ranks. *Sci. Rep.* **2017**, *7*, 2542. [CrossRef]

81. Naim, M.A.; Morillo, J.A.; Sørensen, S.J.; Waleed, A.A.S.; Smidt, H.; Sipkema, D. Host-specific microbial communities in three sympatric North Sea sponges. *FEMS Microbiol. Ecol.* **2014**, *90*, 390–403. [CrossRef]

82. Strehlow, B.W.; Schuster, A.; Francis, W.R.; Canfield, D.E. Metagenomic data for*Halichondria*panicea from Illumina and nanopore sequencing and preliminary genome assemblies for the sponge and two microbial symbionts. *BMC Res. Notes* **2022**, *15*, 135. [CrossRef]

83. Sun, J.F.; Wu, Y.; Yang, B.; Liuc, Y. Chemical constituents of marine sponge *Halichondria* sp. from south China sea. *Chem. Nat. Compd.* **2015**, *51*, 975–977. [CrossRef]

84. Hirata, Y.; Uemura, D. Halichondrins-antitumor polyether macrolides from a marine sponge. *Pure Appl. Chem.* **1986**, *58*, 701–710. [CrossRef]

85. Tachibana, K.; Scheuer, P.J.; Tsukitani, Y.; Kikuchi, H.; Van Engen, D.; Clardy, J.; Gopichand, Y.; Schmitz, F.J. Okadaic acid, a cytotoxic polyether from two marine sponges of the genus Halichondria. *J. Am. Chem. Soc.* **1981**, *103*, 2469–2471. [CrossRef]

86. Amagata, T.; Tanaka, M.; Yamada, T.; Minoura, K.; Numata, A. Gymnastatins and dankastatins, growth inhibitory metabolites of a Gymnascella species from a*Halichondria*sponge. *J. Nat. Prod.* **2008**, *71*, 340–345. [CrossRef]

87. Prego-Faraldo, M.V.; Valdiglesias, V.; Méndez, J.; Eirín-López, J.M. Okadaic acid meet and greet: An insight into detection methods, response strategies and genotoxic effects in marine invertebrates. *Mar. Drugs* **2013**, *11*, 2829–2845. [CrossRef]

88. Fu, L.L.; Zhao, X.Y.; Ji, L.D.; Xu, J. Okadaic acid (OA): Toxicity, detection and detoxification. *Toxicon* **2019**, *160*, 1–7. [CrossRef]

89. Corriere, M.; Soliño, L.; Costa, P.R. Effects of the marine biotoxins okadaic acid and dinophysistoxins on fish. *J. Mar. Sci. Eng.* **2021**, *9*, 293. [CrossRef]

90. Khalaman, V.V.; Sharov, A.N.; Kholodkevich, S.V.; Komendantov, A.Y.; Kuznetsova, T.V. Influence of the White Sea sponge*Halichondria*panicea (Pallas, 1766) on physiological state of the blue mussel Mytilus edulis (Linnaeus, 1758), as evaluated by heart rate characteristics. *J. Evol. Biochem. Physiol.* **2017**, *53*, 225–232. [CrossRef]

91. Pita, L.; Fraune, S.; Hentschel, U. Emerging sponge models of animal-microbe symbioses. *Front. Microbiol.* **2016**, *7*, 2102. [CrossRef]

92. Lemoine, N.; Buell, N.; Hill, A.; Hill, M. Assessing the utility of sponge microbial symbiont communities as models to study global climate change: A case study with*Halichondria*bowerbanki. In *Porifera Research: Biodiversity, Innovation and Sustainability; Série Livros*, 1st ed.; Custódio, M.R., Ed.; Museu Nacional: Rio de Janeiro, Brazil, 2007; Volume 28, pp. 419–425.

93. Frøhlich, H.; Barthel, D. Silica uptake of the marine sponge*Halichondria*panicea in Kiel Bight. *Mar. Biol.* **1997**, *128*, 115–125. [CrossRef]

94. Barthel, D.; Detmer, A. The spermatogenesis of*Halichondria*panicea (Porifera, Demospongiae). *Zoomorphology* **1991**, *110*, 9–15. [CrossRef]

95. Sokolova, A.M.; Pozdnyakov, I.R.; Ereskovsky, A.V.; Karpov, S.A. Kinetid structure in larval and adult stages of the demosponges Haliclona aquaeductus (Haplosclerida) and*Halichondria*panicea (Suberitida). *Zoomorphology* **2019**, *138*, 171–184. [CrossRef]

96. Amano, S. Larval release in response to a light signal by the intertidal sponge*Halichondria*panicea. *Biol. Bull.* **1986**, *171*, 371–378. [CrossRef]

97. Lüskow, F.; Riisgård, H.U.; Solovyeva, V.; Brewer, J.R. Seasonal changes in bacteria and phytoplankton biomass control the condition index of the demosponge*Halichondria*panicea in temperate Danish waters. *Mar. Ecol. Prog. Ser.* **2019**, *608*, 119–132. [CrossRef]

98. Riisgård, H.U.; Thomassen, S.; Jakobsen, H.; Weeks, J.M.; Larsen, P.S. Suspension feeding in marine sponges*Halichondria*panicea and Haliclona urceolus: Effects of temperature on filtration rate and energy cost of pumping. *Mar. Ecol. Prog. Ser.* **1993**, *96*, 177–188. [CrossRef]

99. Koopmans, M.; Martens, D.; Wijffels, R.H. Growth efficiency and carbon balance for the sponge Haliclona oculata. *Mar. Biotechnol.* **2010**, *12*, 340–349. [CrossRef]

100. Riisgård, H.U.; Larsen, P.S. Actual and model-predicted growth of sponges—with a bioenergetic comparison to other filter-feeders. *J. Mar. Sci. Eng.* **2022**, *10*, 607. [CrossRef]

101. Fell, P.E.; Parry, E.H.; Balsamo, A.M. The life histories of sponges in the Mystic and Thames estuaries (Connecticut), with emphasis on larval settlement and postlarval reproduction. *J. Exp. Mar. Biol. Ecol.* **1984**, *78*, 127–141. [CrossRef]

102. Hartman, W.D. Natural history of the marine sponges of southern New England. *Bull. Peabody Mus. Yale* **1958**, *12*, 1–155.

103. Knowlton, A.L.; Highsmith, R.C. Nudibranch-sponge feeding dynamics: Benefits of symbiont-containing sponge to Archidoris montereyensis (Cooper, 1862) and recovery of nudibranch feeding scars by*Halichondria*panicea (Pallas, 1766). *J. Exp. Mar. Biol. Ecol.* **2005**, *327*, 36–46. [CrossRef]

104. Wulff, J. Regeneration of sponges in ecological context: Is regeneration an integral part of life history and morphological strategies? *Integr. Comp. Biol.* **2010**, *50*, 494–505. [CrossRef]

105. Alcolado, P.M. Nuevas especies de esponjas encontradas en Cuba. *Poeyana* **1984**, *271*, 1–22.

106. Wulff, J.L. Parrotfish predation on cryptic sponges of Caribbean coral reefs. *Mar. Biol.* **1997**, *129*, 41–52. [CrossRef]

107. Wulff, J.L. Sponge predators may determine differences in sponge fauna between two sets of mangrove cays, Belize barrier reef. *Atoll Res. Bull.* **2000**, *477*, 251–263. [CrossRef]

108. De Goeij, J.M.; Van Oevelen, D.; Vermeij, M.J.; Osinga, R.; Middelburg, J.J.; De Goeij, A.F.; Admiraal, W. Surviving in a marine desert: The sponge loop retains resources within coral reefs. *Science* **2013**, *342*, 108–110. [CrossRef]

109. Barthel, D.; Wolfrath, B. Tissue sloughing in the sponge*Halichondria*panicea: A fouling organism prevents being fouled. *Oecologia* **1989**, *78*, 357–360. [CrossRef]

110. Hansen, I.V.; Weeks, J.M.; Depledge, M.H. Accumulation of copper, zinc, cadmium and chromium by the marine sponge*Halichondria*panicea Pallas and the implications for biomonitoring. *Mar. Pollut. Bull.* **1995**, *31*, 133–138. [CrossRef]

111. Langenbruch, P.F.; Scalera-Liaci, L. *Structure of Choanocyte Chambers in Haplosclerid Sponges*; Smithsonian Institution Press: Washington, DC, USA; Woods Hole, MA, USA, 1985; pp. 245–251.

112. Haeckel, E. XXVII. On the Calcispongiae, their position in the animal kingdom, and their relation to the theory of descendence. *J. Nat. Hist.* **1873**, *11*, 241–262.

113. Reiswig, H.M. The aquiferous systems of three marine Demospongiae. *J. Morphol.* **1975**, *145*, 493–502. [CrossRef]

114. Fry, W.G. The sponge as a population: A biometric approach. *Symp. Zool. Soc. Lond.* **1970**, *25*, 135–162.

115. Fry, W.G. Taxonomy, the individual and the sponge. In *Biology and Systematics of Colonial Organisms: Proceedings of an International Symposium Held at the University of Durham*; Published for the Systematics Association; Academic Press: London, UK, 1979; Volume 11, pp. 39–80.

116. Riisgård, H.U.; Larsen, P.S. Filtration rates and scaling in demosponges. *J. Mar. Sci. Eng.* **2022**, *10*, 643. [CrossRef]
117. Larsen, P.S.; Riisgård, H.U. The sponge pump. *J. Theor. Biol.* **1994**, *168*, 53–63. [CrossRef]
118. Weissenfels, N. The filtration apparatus for food collection in freshwater sponges (Porifera, Spongillidae). *Zoomorphology* **1992**, *112*, 51–55. [CrossRef]
119. Mah, J.L.; Christensen-Dalsgaard, K.K.; Leys, S.P. Choanoflagellate and choanocyte collar-flagellar systems and the assumption of homology. *Evol. Dev.* **2014**, *16*, 25–37. [CrossRef]
120. Suarez, P.A.; Leys, S.P. The sponge pump as a morphological character in the fossil record. *Paleobiology* **2022**, *48*, 446–461. [CrossRef]
121. Evans, C.W. The ultrastructure of larvae from the marine sponge*Halichondria*moorei Bergquist (Porifera, Demospongiae). *Cah. Biol. Mar.* **1997**, *18*, 427–433.
122. Harris, A.K. Cell motility and the problem of anatomical homeostasis. *J. Cell. Sci.* **1987**, *8*, 121–140. [CrossRef]
123. Gaino, E.; Burlando, B. Sponge cell motility: A model system for the study of morphogenetic processes. *Ital. J. Zool.* **1990**, *57*, 109–118. [CrossRef]
124. Bond, C. Continuous cell movements rearrange anatomical structures in intact sponges. *J. Exp. Zool.* **1992**, *263*, 284–302. [CrossRef]
125. Lavrov, A.I.; Kosevich, I.A. Sponge cell reaggregation: Cellular structure and morphogenetic potencies of multicellular aggregates. *J. Exp. Zool. A. Ecol. Genet. Physiol.* **2016**, *325*, 158–177. [CrossRef]
126. Saito, Y. Self and nonself recognition in a marine sponge,*Halichondria*japonica (Demospongiae). *Zool. Sci.* **2013**, *30*, 651–657. [CrossRef]
127. Storr, J.F. Field observations of sponge reactions as related to their ecology. In *Aspects of Sponge Biology*; Harrison, F.W., Cowden, R.R., Eds.; Academic Press Inc.: New York, NY, USA, 1976; pp. 277–282.
128. Reiswig, H.M. In situ pumping activities of tropical Demospongiae. *Mar. Biol.* **1971**, *9*, 38–50. [CrossRef]
129. Nickel, M. Kinetics and rhythm of body contractions in the sponge Tethya wilhelma (Porifera: Demospongiae). *J. Exp. Biol.* **2004**, *207*, 4515–4524. [CrossRef]
130. Nickel, M.; Scheer, C.; Hammel, J.U.; Herzen, J.; Beckmann, F. The contractile sponge epithelium sensu lato—Body contraction of the demosponge Tethya wilhelma is mediated by the pinacoderm. *J. Exp. Biol.* **2011**, *214*, 1692–1698. [CrossRef]
131. Elliott, G.R.; Leys, S.P. Coordinated contractions effectively expel water from the aquiferous system of a freshwater sponge. *J. Exp. Biol.* **2007**, *210*, 3736–3748. [CrossRef]
132. Kumala, L.; Canfield, D.E. Contraction dynamics and respiration of small single-osculum explants of the demosponge *Halichondria* panicea. *Front. Mar. Sci.* **2018**, *5*, 410. [CrossRef]
133. Kumala, L.; Larsen, M.; Glud, R.N.; Canfield, D.E. Spatial and temporal anoxia in single-osculum*Halichondria*panicea demosponge explants studied with planar optodes. *Mar. Biol.* **2021**, *168*, 173. [CrossRef]
134. Leys, S.P. Elements of a 'nervous system' in sponges. *J. Exp. Biol.* **2015**, *218*, 581–591. [CrossRef]
135. Gerrodette, T.; Flechsig, A.O. Sediment-induced reduction in the pumping rate of the tropical sponge Verongia lacunosa. *Mar. Biol.* **1979**, *55*, 103–110. [CrossRef]
136. Bell, J.J.; McGrath, E.; Biggerstaff, A.; Bates, T.; Bennett, H.; Marlow, J.; Shaffer, M. Sediment impacts on marine sponges. *Mar. Pollut. Bull.* **2015**, *94*, 5–13. [CrossRef]
137. Nickel, M. Evolutionary emergence of synaptic nervous systems: What can we learn from the non-synaptic, nerveless Porifera? *Invertebr. Biol.* **2010**, *129*, 1–16. [CrossRef]
138. Jones, W.C. Is there a nervous system in sponges? *Biol. Rev.* **1962**, *37*, 1–47. [CrossRef]
139. Ludeman, D.A.; Farrar, N.; Riesgo, A.; Paps, J.; Leys, S.P. Evolutionary origins of sensation in metazoans: Functional evidence for a new sensory organ in sponges. *BMC Evol. Biol.* **2014**, *14*, 3. [CrossRef]
140. Leys, S.P.; Nichols, S.A.; Adams, E.D. Epithelia and integration in sponges. *Integr. Comp. Biol.* **2009**, *49*, 167–177. [CrossRef]
141. Bagby, R.M. The fine structure of myocytes in the sponges Microciona prolifera (Ellis and Solander) and Tedania ignis (Duchassaing and Michelotti). *J. Morphol.* **1966**, *118*, 167–181. [CrossRef]
142. Hammel, J.U.; Nickel, M. A new flow-regulating cell type in the demosponge Tethya wilhelma—Functional cellular anatomy of a leuconoid canal system. *PLoS ONE* **2014**, *9*, e113153. [CrossRef]
143. Kanzawa, N.; Takano-Ohmuro, H.; Maruyama, K. Isolation and characterization of sea sponge myosin. *Zool. Sci.* **1995**, *12*, 765–769. [CrossRef]
144. De Ceccatty, M.P. Coordination in sponges. The foundations of integration. *Am. Zool.* **1974**, *14*, 895–903. [CrossRef]
145. Perovic, S.; Krasko, A.; Prokic, I.; Müller, I.M.; Müller, W.E. Origin of neuronal-like receptors in Metazoa: Cloning of a metabotropic glutamate/GABA-like receptor from the marine sponge Geodia cydonium. *Cell Tissue Res.* **1999**, *296*, 395–404. [CrossRef]
146. Lieberkühn, N. *Über Bewegungserscheinungen der Zellen*; Band IX, N.G., Ed.; Elwert'sche Universitätsbuchhandlung: Marburg/Leipzig, Germany, 1870; pp. 9–22.
147. Galtsoff, P.S. The amoeboid movement of dissociated sponge cells. *Biol. Bull.* **1923**, *45*, 153–161. [CrossRef]
148. Loewenstein, W.R. On the genesis of cellular communication. *Devel. Biol.* **1967**, *15*, 503–520. [CrossRef]

Journal of
Marine Science and Engineering

Review

Size-Specific Growth of Filter-Feeding Marine Invertebrates

Poul S. Larsen [1],* and Hans Ulrik Riisgård [2]

[1] Department of Mechanical Engineering, Technical University of Denmark, 2800 Kongens Lyngby, Denmark
[2] Marine Biological Research Centre, University of Southern Denmark, 5300 Kerteminde, Denmark
* Correspondence: psl@mek.dtu.dk

Abstract: Filter-feeding invertebrates are found in almost all of the animal classes that are represented in the sea, where they are the necessary links between suspended food particles (phytoplankton and free-living bacteria) and the higher trophic levels in the food chains. Their common challenge is to grow on the dilute concentrations of food particles. In this review, we consider examples of sponges, jellyfish, bryozoans, polychaetes, copepods, bivalves, and ascideans. We examine their growth with the aid of a simple bioenergetic growth model for size-specific growth, i.e., in terms of dry weight (W), $\mu = (1/W)\,dW/dt = aW^b$, which is based on the power functions for rates of filtration ($F \approx W^{b1}$) and respiration ($R \approx W^{b2}$). Our theory is that the exponents have (during the evolution) become near equal ($b_1 \approx b_2$), depending on the species, the stage of ontogeny, and their adaptation to the living site. Much of the compiled data support this theory and show that the size-specific rate of growth (excluding spawning and the terminal phase) may be constant ($b = 0$) or decreasing with size ($b < 0$). This corresponds to the growth rate that is exponential or a power function of time; however, with no general trend to follow a suggested 3/4 law of growth. Many features are common to filter-feeding invertebrates, but modularity applies only to bryozoans and sponges, implying exponential growth, which is probably a rather unique feature among the herein examined filter feeders, although the growth may be near exponential in the early ontogenetic stages of mussels, for example.

Keywords: filtration; respiration; bioenergetic growth model; exponential growth; power function growth

Citation: Larsen, P.S.; Riisgård, H.U. Size-Specific Growth of Filter-Feeding Marine Invertebrates. *J. Mar. Sci. Eng.* **2022**, *10*, 1226. https://doi.org/10.3390/jmse10091226

Academic Editor: Azizur Rahman

Received: 22 July 2022
Accepted: 31 August 2022
Published: 2 September 2022

Publisher's Note: MDPI stays neutral with regard to jurisdictional claims in published maps and institutional affiliations.

1. Introduction and Growth Model

Filter-feeding (or suspension-feeding) marine invertebrates are important animals in the food chains of the sea [1,2]. They trap food particles, such as phytoplankton and bacteria, from a feeding current that is created by their own pumping device or by the ambient, and the mechanisms that are used in order to capture and transport the particles to be ingested reflect various secondary adaptations to filter feeding among species [3]. The rate of growth of individuals is of interest in estimating the population grazing impact at a particular site, hence its ecological significance. However, the growth rates may also be of commercial interest, e.g., in mussel farming [4].

Here, we focus on the somatic growth of filter-feeding marine invertebrates under favorable conditions and exclude the release of biomass that is associated with spawning and terminal growth. We consider sponges (*Halichondria panicea*), jellyfish (*Aurelia aurita*), bryozoans (*Electra pilosa* and *Celleporella hyalina*), polychaetes (*Nereis diversicolor* and *Sabella spallanzanii*), crustaceans (calanoid copepods), bivalves (*Mytilus edulis*), and ascideans (*Ciona intestinalis*). Among these, only *N. diversicolor* is a facultative filter feeder that may switch to surface deposit feeding or scavenging.

Filter feeding in all marine invertebrates is a secondary adaptation. The blue mussel, *Mytilus edulis*, is a well-known example where the gills have become the water-pumping and particle-capturing organ, while the original function as gills has become superfluous [1,5]. In crustaceans, the secondary adaptation to filter feeding has evolved independently in many groups, and often filter feeding is only one of several feeding methods that are adopted by one species. A common feature is that the filter-feeding process is true sieving, where the mesh size of the filter determines the size of the captured food particles [6]. However, a secondary adaptation to filter feeding must have involved the development of a filter pump that can cover the need for food energy in order to cover the respiration requirements. When the respiration (R) increases with increasing body mass, the filtration (F) must necessarily follow and, therefore, our theory is that the exponents in the equation $F = a_1 W^{b1}$ and $R = a_2 W^{b2}$ have (during their evolution) become near equal ($b_1 \approx b_2$), depending on the species, the stage of ontogeny, and their adaptation to the living site.

Here, *Mytilus edulis* is a good example as $b_1 \approx 1$ in very small juvenile mussels but decreases to $b_1 = 0.66$ in larger mussels [7]. This makes the bioenergetic growth model [8,9] particularly simple, expressing the growth rate as $G = aW^{b1}$ and the weight-specific growth rate as follows:

$$\mu \equiv (1/W)\, dW/dt = aW^b;\ b = b_1 - 1, \tag{1}$$

where the coefficient $a = (C \times AE \times a_1 - a_2)/a_0$ depends on the food concentration (C), the assimilation efficiency (AE), and the cost of growth (R_g/G), i.e., the metabolic cost of synthesizing new biomass, being the equivalent fraction of $a_0 - 1$ of the growth itself. Because the proportion of the biomass with a low metabolism (e.g., lipid and glycogen store, gonads) may increase with the body size, the weight-specific respiration may concurrently decrease due to a negative exponent ($b = b_1 - 1$).

Depending on the species that is investigated, the growth parameter of the dry body weight (W) could be the tissue volume (V) (sponges), the area (A) or number (N) of zooids in an encrusting colony (bryozoans), or a characteristic length (ascidians). Furthermore, it is of interest to note that the growth function $W(t)$ may take two specific forms that are simply related to the specific growth rate, i.e., exponential growth, as follows:

$$W(t) \sim \exp(at),\ \mu = a, \tag{2}$$

and power function growth as follows:

$$W(t) \sim t^d,\ \mu \sim W^{-1/d}. \tag{3}$$

If $b_1 \approx b_2 \approx 1$, hence $b = 0$ in a filter-feeding animal, then $\mu = a$, and the growth is exponential, but otherwise the weight-specific growth rate will decrease with increasing body size.

The present theory of near-equal b-exponents forms the underlying basis of this review, where we consider examples of filter-feeding invertebrates and examine if their growth functions are exponential or power functions of time. The b-exponents of rates of filtration and respiration that have been found for the examined filter feeders are compiled in Table 1, from which it appears that, in many cases, $b_1 \approx b_2$ in support of the theory, and the typical growth rates are listed in Table 2.

Table 1. b-exponents of rates of filtration b_1 (in $F = a_1 W^{b_1}$) and respiration b_2 (in $R = a_2 W^{b_2}$) of some filter-feeding invertebrates. *Halichondria panicea*, "small" = single osculum sponge explant; "large" = multi-oscula sponge.

Filter Feeder	b_1		b_2	
Sponges *Halichondria panicea*, small	~2/3	[10]	~2/3	[10]
Halichondria panicea, large	~1.0	[11]	~1.0	[11]
Jellyfish *Aurelia aurita*	0.78	[12]	0.86	[13]
Bryozoans *Electra pilosa*				
Polychaetes *Nereis diversicolor*	1.0	[14]	1.2	[15]
Crustaceans calanoid copepods	0.84	[16]	0.78	[16]
Bivalves *Mytilus edulis*, W < 10 mg	0.887	[7]	1.03	[7]
Mytilus edulis, W > 10 mg	0.663	[7]	0.66	[7]
Ascidians *Ciona intestinalis*	0.68	[17]	0.831	[18]

Table 2. Specific growth rates of some filter-feeding marine invertebrates. Growth is measured as the change in size of the following: (W) = dry weight, (A) = colony area, (N) = number of zooids, (L) = body length. Cost of growth = R_g/G.

Specimen	μ (% d^{-1})		Comment	Cost of Growth	
Sponges *Halichondria panicea*	0.6 to 1.18 (W)	[11,19]	max = 4% d^{-1} [11]	1.39	[11]
Jellyfish *Aurelia aurita*	24.4 (W) to 6.28 (W)	[20,21]	2.2 mg to 115–1887 mg	0.2	[19]
Bryozoans *Electra pilosa*	9 (A), 11 (N)	[22]			
Celleporella hyalina	10 (A), 12 (N)	[22]			
5 cheilostome species	12 (N) to 4 (N)	[23]	100 to 600 zooids		
Polychaetes *Nereis diversicolor*	3.1 to 3.9 (W)	[24]	fed on algea	0.08	[15]
Nereis diversicolor	7 (W)	[15]	fed on scrimp meat	0.26	[15]
Sabella spallanzanii	0.74~1 (W)	[25]			
Crustaceans calanoid copepods	4 to 135 (W)	[26]	L = 50 to 80 μm		
Bivalves *Mytilus edulis*	7.8 to 1.6 (W)	[19]	W = 0.01 g to 1 g	0.12	[19]
Ascidians *Ciona intestinalis*	1.4 to 3 (L), 7.7 (W)	[27,28]		0.21 − 0.23	[27]

2. Sponges

Sponges are one of the earliest evolved and simplest groups of animals [29]. They are multicellular filter feeders that actively pump volumes of water, which are equivalent to about six times their body volume per minute, through their canal system by means of flagellated choanocytes that are arranged in choanocyte chambers. They feed on free-living bacteria that are trapped on the array of microvilli of the collars of their choanocytes, and on larger phytoplankton cells that are drawn into the inhalant canal system, where they are captured and phagocytosed [30–35]. The water enters the sponge body through numerous small inhalant openings (ostia) and passes through incurrent canals, the choanocyte chambers, and through excurrent canals to be expelled as a jet through an exhalant opening (osculum). The water flow also ensures a supply of oxygen for respiration via diffusive oxygen uptake [36].

The experimental results that were summarized in [19] for the demosponge *Halichondria panicea* show the exponential growth at rates of μ = 0.6% to 1.18% d^{-1}, and the maintenance food concentration (at no growth) was estimated for a given specimen to be C_m = 25.1 μg C L^{-1}. The highest reported growth rate of *H. panicea* of 4% d^{-1} [11] corresponds to an available total carbon biomass (TCB) of $C = 8\,C_m$, i.e., eight times that of the maintenance food concentration. The growth rate may thus be expected to increase linearly with C to a maximal value that is not exceeded, irrespective of how high C becomes.

In fair agreement with the observed exponential growth of *Halichondria panicea*, [11] found that exponents of the power functions of the filtration and respiration rates were b_1 = 0.914 and b_2 = 0.927, while [37] found that $b_1 \approx b_2 \approx 1$ for three larger tropical marine sponges. It is, therefore, of interest to see how the filtration rates of sponges depend on their size, which was summarized in [38] and approximated by the power function

$F/V = a_3 V^{b3}$, because this implies that $b_3 = b_1 - 1 = b$ of Equation (1), assuming that the dry weight was proportional to the volume (V). Many values of the exponents b_3, which have been reported by various authors that were cited in [38], suggest that $b \approx 0$, implying exponential growth. However, in other cases, $b < -0.2$, and as low as -0.7, implying power function growth, which, however, is subject to the uncertainty of the assumption $V \sim W$ for larger sponges.

Related to these considerations of volume-specific filtration rate is a recent discussion of its possible scaling to the size of demosponges. Here, according to [10], demosponges are modular filter feeders where the early stages of single-osculum aquiferous modules have volume-specific pumping rates that scale as $F/V = a_3 V^{b3}$, $b_3 \approx -1/3$ (hence $b_2 = 1 + b_3 = 2/3$, Table 1), as measured for small explants of *Halichondria panicea* [39,40]. Such modules only grow to a certain size [41,42], such that new modules will be formed for a growing sponge. Larger multi-oscular sponges consist mainly of single-oscular modules that have reached their maximal size, hence $b_3 \approx 0$, as noted for some species that were mentioned in [38]. A similar situation may be expected for large pseudo-oscular sponges (e.g., *Xestospongia muta* [43,44]), provided that the volumes were taken to be that of structural tissue and not the total volume that includes the large atrium volume of water. In summary, these considerations and $b_1 \approx b_2$ suggest that $b \approx -1/3$ for growing single-osculum demosponges and $b \approx 0$ for multi-oscula sponges, according to Equation (1).

3. Jellyfish

Medusae of the common jellyfish *Aurelia aurita* are voracious predators that filter feed on zooplankton. They occur in many coastal ecosystems where they can be very abundant and so exert a considerable predation impact on the zooplankton; thus, they play an important role as a key organism in the ecosystem [45]. The predation impact can be evaluated when the population density and the individual clearance rate of the jellyfish are known. The weight-specific growth rate μ can be determined from the time interval collection, measuring their umbrella diameters, and converting that to dry weight.

Controlled laboratory experiments have used brine shrimp *Artemia* sp. as prey, for which the retention efficiency (60%) [46] is much higher and well defined than for natural prey, such as copepods, probably due to a lack of escape behavior. A study [20] found a typical growth rate of $\mu = 10\%$ d^{-1} and the exponent $b = -0.2$ in Equation (1) for fully fed larger specimens ($W = 154$ mg, $d = 56.9$ mm umbrella diameter) with a diet of five *Artemia* L^{-1}. A value of $b = -0.2$ is in fair agreement with the values $b_1 = 0.78$ and $b_2 = 0.86$, which were determined earlier [12,13]. For smaller specimens ($W = 2.2$ mg, $d = 12.7$ mm), the growth rate was $\mu = 24.4\%$ d^{-1}, which was in good agreement with the growth model of Equation (1) for $b = -0.2$. Similar experiments [21] showed $b = -0.24$ and a mean growth rate of $\mu = 6.28\%$ d^{-1} over a size range of 115 to 1887 mg. According to the growth model of Equation (1), μ increases from zero (at the maintenance concentration, 1.23 *Artemia* L^{-1}) linearly with the prey concentration, but only to a maximal value (at about five times that of the maintenance concentration), above which there is no further increase, and such maximal values depend on the size, see [45].

The growth rates from the field data, as indicated by the values of $b = -0.43$ to -0.84, which are summarized in [45], decrease much more with increasing size than those from the laboratory, which were close to the suggested theoretical value of $b < -0.2$. This difference can be explained by the lower retention efficiency and the fluctuating prey concentrations.

4. Bryozoans

Bryozoans are sessile, colonial filter feeders that mainly feed on phytoplankton that are drawn in with the flow into the tentacle crown (lophophore) by the cilia on the tentacles, which act as pumps and help to retain and transfer the prey to the mouth. In encrusting bryozoans, the filtered water flows under the lophophore canopy in order to escape at the edges of the colony or as jets from 'chimneys' inside of the colony, which, as a biomixing process, may help to prevent re-filtration. Encrusting bryozoans may form colonies of essentially identical zooids, which have the same filtration rates, and hence may be considered to be modules of an 'organism', the colony [19]. The growth occurs by adding zooid modules along the periphery and may be measured by the increase with time of the area (A) or the number (N) of zooids in the colony. By denoting the area of one zooid as A_z, the relation for incremental growth may be written as $dA = A_z\,dN$. If A_z remains constant during the growth, the specific growth rates - μ_A and μ_N would be equal, but if A_z decreased with growth, we expect $\mu_A < \mu_N$.

The growth rates of colonies of *Electra pilosa* and *Celleporella hyalina* that were placed on microscope slides in both field and laboratory tests were measured in [22]. For the laboratory tests at an algal concentration of about 5000 *Rhodomonas* cells mL^{-1}, representing well-fed conditions, the growth was found to be exponential, with specific growth rates for *E. pilosa* being $\mu_A = 0.09 \pm 0.02$ d^{-1} and $\mu_N = 0.11 \pm 0.02$ d^{-1}, and for *C. hyaline* being $\mu_A = 0.10 \pm 0.01$ d^{-1} and $\mu_N = 0.12 \pm 0.01$ d^{-1}. From the ratio of number-to-area, the density was in the range of five to eight zooids per mm^2. The higher growth rate, which was based on the number of zooids, implies an increasing density of zooids with increasing colony size. The field data for *C. hyaline* (genotype H) showed similar exponential growth rates but had slightly smaller specific growth rates, which was likely due to the larger zooid size and the lower algal concentration (<1500 cells mL^{-1}). The absence of the influence of the current velocity was ascribed to the fact that the thin layer of encrusting bryozoans is well within the low velocity viscous sublayer.

Among the 23 laboratory and field data sets for different species [22] the average specific growth rates were $\mu_A \sim 0.09$ to 0.14 d^{-1}, while some lower and higher values could be caused by high algal concentrations (> 5000 cells mL^{-1}). The low values (0.06–0.08 d^{-1}) were observed for the natural colonies that were feeding on macroalgae. This study gives the orders of magnitude of specific growth rates and indicates the growth rate to be exponential.

Ref. [23] analyzed the datasets from [47] for the growth rate of five fouling marine cheilostome colony species and found the growth rate to follow power functions ($N = a_p\,t^d$), with exponent $d = 2.266 \pm 0.214$ d^{-1} as the average of 10 groups. Apparently, the rate of asexual zooid replication increases with the colony size in many bryozoan species, hence the switch from exponential growth to power function growth. Thus, according to Equation (3), the number-specific growth rate ($\mu_N = (1/N)\,dN/dt = d\,a_p^{1/d}\,N^{-1/d}$) decreases with increasing colony size as $\mu_N \sim N^{-0.44}$, according to the data for July to August of [23] from $\mu_N = 0.12$ d^{-1} at $N = 100$ zooids to $\mu_N = 0.041$ d^{-1} at $N = 600$ zooids. Interestingly, the first value of the specific growth rate is close to that observed in [22] for a growth up to a size of about 100 zooids, however, it became smaller in the larger colonies. This trend is similar to that observed for the blue mussel and may be ascribed to a change in the composition of the biomass. Although it is not supported by explicit data, it is possible that the rates of both filtration and respiration may be proportional to the number of zooids in a smaller growing colony of modules of developed zooids, which would imply $b_1 = b_2 \approx 1$, hence $b = 0$ in the bioenergetics model of Equation (1), implying exponential growth, as found in [22]. However, for the larger colonies, the growth follows a power function, as found in [23].

5. Polychaetes

The facultative filter-feeding polychaete *Nereis diversicolor* can feed by pumping water through a mucus net-bag that is attached to the entrance of its U-shaped burrow in the sediment. The retained food particles in the net are then ingested with the rolled-up net. Switching from surface deposit feeding to filter feeding in *N. diversicolor* is an adaptation that is useful for the near-bottom dweller in shallow waters, where the concentration of suspended food particles varies widely. Filter feeding, in place of surface deposit feeding or scavenging/predation, is preferred when there is a sufficient concentration of suspended algal cells; however, the full growth potential of the polychaete is not always achieved near to the bottom due to food depletion in the absence of efficient vertical mixing. Field growth studies [24] were therefore carried out with the worms placed in glass tubes at 15 cm above the bottom. This showed the weight-specific growth rates increasing to $\mu = 3.9\%\,d^{-1}$ for an increasing concentration of Chl *a* of available algal cells, which is comparable to the results from the laboratory studies showing $\mu = 3.1\%\,d^{-1}$, and the cost of growth was estimated to be only 8% of the total growth. Later laboratory studies [15] showed a growth rate of $\mu = 3.0\%\,d^{-1}$ for a *Rhodomonas* algal diet but $\mu = 7\%\,d^{-1}$ for a shrimp meat diet, and in both cases the weight-specific respiration (R/W) increased linearly with the specific-growth rate μ. This increase, which is also called 'specific dynamic action', indicated an energy cost of growth that was equivalent to 26% of the total growth, which is similar to some of the other filter feeders that are shown in Table 2. These studies show the effect of the nutritional value of the diet, from algae to shrimp meat, apparently raising both the weight-specific growth rate and cost of growth by the same factor of approximately two.

The obligate filter-feeding polychaete *Sabella spallanzanii* lives in a tube from which it extends its feeding organ that consists of a filament crown with closely spaced pinnules whose rows of compound latero-frontal cilia pump the water through the space between the pinnules and retain the food particles. It lives in patches and [25] recorded a density of 150 ind. m^{-2} and observed the growth in terms of the mean length of the tubes from L = 10 to 18 cm during the period from April to August of 1992, implying a mean length-specific growth rate of $\mu_L = 100 \times \ln(18/10)/120 = 0.49\%\,d^{-1}$. They also reported an increase in the biomass (that we assume is proportional to the dry mass of the animals) from $60\,g\,m^{-2}$ to $75\,g\,m^{-2}$ for one month, which may be interpreted as a mean weight-specific growth rate over that period of $\mu = (1/W)\,dW/dt = 100 \times \ln(75/60)/30 = 0.74\%\,d^{-1}$. The relationship between these specific growth rates agrees with the relationship between the biomass $W(g)$ and the total worm length $L(cm)$ that is given approximately as $W = 0.0021\,L^2 - 0.0098\,L$, which is derived from [25].

6. Copepods

Raptorial feeding is probably primary in copepods, whereas filter feeding is a specialized condition that has been developed within the order of calanoid copepods. For example, the calanoid copepod *Acartia tonsa* uses ambush feeding when it slowly sinks through the water with extended antennae that perceive motile prey, or it uses filter feeding by generating a feeding current with which phytoplankton cells are captured by a filter that is formed by setae on its appendages. Copepods need to consume a large number of phytoplankton per day in order to cover their nutritional needs. They are characterized by a relatively fast growth rate [48] and page 428 of [6].

The growth rates of three calanoid copepods were determined in [49]. The growth rates were low, particularly during the summer. The specific growth rates of the copepodites were moderately high for *Eurytemora affinis* in the spring as follows: 23% and 15% d^{-1} for the early and late stages, respectively, and were low for *Pseudodiaptomus forbesi* in the summer as follows: 15% and 3% d^{-1}, respectively (not shown in Table 2).

Small copepods grow fast at high temperatures accompanied with an ample food supply, as shown in [26], which is a study from the East China Sea. Thus, the weight-specific growth rates ranged from 4% to 135% d^{-1} in the 50 to 80 µm size fraction, and from 1% to 79% d^{-1} in the 100 to 150 µm size fraction, showing that the growth rates were positively related to the temperature and were negatively related to the body size. The strong size dependence could imply a power function growth with a small negative *b*-value, as indicated by the exponent *b* of Equation (1), increasing from $b = -0.88$ to $b = -0.32$ for the two size fractions [26]. However, according to Table 1 of [16] for calanoid copepods, the exponents of the power functions of weight-specific rates of clearance and respiration take the following values: $b_1 = -0.16 + 1.0 = 0.84$, $b_2 = -0.22 + 1.0 = 0.78$, or $b_1 \approx b_2 \approx 0.8$, hence $b = -0.2$ for the weight-specific growth rate of Equation (1). The value of $b = -0.2$ may be compared with the somewhat higher value $b = -0.06$ in Table 1 of [16], both of which seem to suggest power-function growth.

7. Bivalves

Most filter-feeding bivalve species have flat gills and employ essentially the same feeding mechanism. There is a wide diversity of gill types, and two different principles are used for food capture, depending on the presence or absence of latero-frontal cirri (see reviews [3,50]). We can direct attention to the blue mussel, *Mytilus edulis*, because it is the most abundant and most studied, which is partly because of its role in commercial aqua farming. The pump that draws in the water through an inhalant mantel opening into the mantel cavity, through W-shaped gills, and discharges it through an exhalant siphon as a jet, is the lateral cilia on the sides of the gill filaments. The separation of the food particles from the pumped current and their retention is handled by the latero-frontal cirri, ensuring a near 100% retention of particles above about 4 μm in size, which includes most phytoplankton. Filter feeding is a secondary adaptation where the gills have become greatly enlarged—far more than what is needed for respiration—which may be suggested to be 'evolutionary adapted' to the prevailing (often low) level of phytoplankton in the surrounding water [5].

The growth rates may change with the increasing size of the specimens. For *Mytilus edulis*, there is a shift at a dry weight of around $W = 10$ mg, corresponding to the shell length $L = 10$ mm, moving from the smaller post-metamorphic mussels of near-exponential growth to the larger juvenile specimens of power function growth. This is reflected by the following values of exponents for the rates of filtration and respiration, which are summarized in [51]: $W < 10$ mg: $b_1 = 1.03$ and $b_2 = 0.887$, or $b_1 \approx b_2 \approx 0.9$ and $W > 10$ mg: $b_1 \approx b_2 \approx 0.66$ (see Table 1).

In order to appreciate the influence of the food concentration on the weight-specific growth rate, Equation (1), with appropriate numerical coefficients (see [9]), takes the following form:

$$\mu \equiv (0.871 \times C - 0.986)\, W^{-0.34};\ W > 0.01\ \mathrm{g}, \tag{4}$$

where units are μ (% d^{-1}), C (μg Chl *a* L^{-1}), and W(g). The measured growth rates in the field at $C = 2.8$ μg to 3.6 μg Chl *a* L^{-1} [4] are in good agreement with Equation (4), which also shows that starvation should occur at $C \leq 0.986/0.871 = 1.13$ μg Chl *a* L^{-1}, while saturation has been found to occur at $C \approx 8$ μg Chl *a* L^{-1} [52]. For a typical value of $C \approx 3$ μg Chl *a* L^{-1}, the growth rate decreases as $\mu = 7.8\%$ to 1.6% d^{-1} for a dry weight increase of $W = 0.01$ to 1.0 g.

The growth rates, in terms of the shell length, may be obtained from Equation (4) by the use of the allometric relation $W(\mu g) = 2.15\, L(\mathrm{mm})^{3.40}$, which shows that $W = 0.01$ g to 1.0 g corresponds to $L = 12$ mm to 46 mm. Furthermore, Equation (1) may be integrated in time in order to show the time that it takes to grow a mussel of a certain size. Finally, the bioenergetic model of Equation (1) has been extended to include the effects of low temperature and low food concentration [53].

8. Ascidians

The benthic ascidian *Ciona intestinalis* may often exert a significant grazing impact because of its dense populations in shallow waters. It retains food particles, including free-living bacteria of < 2 μm, on its mucous net with 80% to 100% efficiency [54]. It is characterized by rapid growth, early maturation, and a high reproductive output.

The rate of oxygen consumption was found in [18] to be dependent on the dry weight to the power $b_2 = 0.831$, which would suggest power function growth according the bioenergetic model for $b_1 \approx b_2$ as $\mu \sim W^{-0.17}$. It is important to note, however, the lower value of $b_1 = 0.68$ in Table 1. However, growth can be exponential until it reaches a body length of 10 mm [55], which could suggest a switch from $\mu \sim$ constant for smaller specimens to a decreasing μ for large specimens.

The growth, in terms of length, has been reported to be 10 mm to 20 mm per month [56], 0.26 mm to 0.76 mm in diameter in seven days [57], or 0.7% d^{-1}, increasing to a maximum of 1.4% to 3% d^{-1} [27]. The weight-specific growth rate increases with increasing temperature and algal concentration to about 7.7% d^{-1} [17,56], and it appears that a condition index, which is defined as the ratio of the dry weight of the soft parts to the total dry weight, is a good indicator of growth as it increases linearly with the weight-specific growth rate up to 8% d^{-1} [27,28].

9. Adaptation to Filter Feeding

The present theory of near-equal b-exponents of filtration and respiration laws, which forms the underlying basis of the present review, is supported by the examples of filter-feeding invertebrates that are compiled in Table 1, from which it appears that $b_1 \approx b_2$. Furthermore, Figures 1A and 2A of [16] confirm this theory by showing that, for a large number of marine pelagic animals, $b_1 \approx b_2 \approx 1$ on the average, but with considerable scatter, and perhaps not $b_1 \approx b_2 \approx 3/4$, as was suggested by the authors. For a specific animal, however, the values of the exponents often take values that are different from 1 and 3/4, as seen from Table 1 of [16] and Table 1 herein, which suggests that there may not be a universal 3/4-law.

All obligate filter-feeding invertebrates (apart from predatory jellyfish) face the same challenge of growing on a thin soup of bacteria and phytoplankton. This fact suggests the existence of some common traits among these animals, which have been identified in [19], as at least the following two features: the magnitude of the filtration–respiration ratio (F/R) and the oxygen extraction efficiency (EE).

One interpretation of the F/R ratio is its relation to food uptake. Here, the estimated $F/R = 22.9$ L of water per mL of O_2 consumed by a specimen of the demosponge *Halichondria panicea* [19] is well above the minimum value of $F/R_m = 10$ L of water (mL O_2)$^{-1}$ [58] for a true filter feeder. Here, $F/R_m = 10$, being the maintenance value $= 20/(0.8 \times 2.5)$, 20 J, i.e., the metabolic equivalent of 1 mL of O_2 and filtering water with a phytoplankton concentration of 1.5 µg Chl a L^{-1} ($= 2.5$ J L^{-1}) with 100% retention and an 80% assimilation efficiency. The minimum value of $F/R_m = 10$ L water (mL O_2)$^{-1}$ was based on a phytoplankton concentration of 1.5 µg Chl a L^{-1} ($= 1.5 \times 40$) $= 60$ µg C L^{-1}. However, sponges also feed on free-living heterotrophic bacteria, cyanobacteria, and other small (0.2−2 µm) picoplankton, as shown in [59], which measured $F/R = 22.9$ L of water per mL of O_2 for *H. panicea*, which appears to agree with the reported carbon concentrations of 40 µg to 200 µg C L^{-1}. Other values for demosponges, which were cited in [19], include 22.8 for *Mycale* sp. and 19.6 for *Tethya crypte*, which are of a similar magnitude. However, the much smaller value of 4.1 for *Verongia gigante* signifies a species that is not a true filter feeder, because it "consists of a tripartite community: sponge-bacteria-polychaete" [37]. [58] shows the $F/R > 10$ L of water per mL of O_2 values for most of the species that are mentioned of the following taxonomic groups: sponges, bryozoans, copepods, polychaetes, bivalves, ascidians, and lancelets. Thus, the F/R ratio is an indicator of the ability of a filter-feeding invertebrate to survive on a pure diet of phytoplankton of 1.5 Chl a L^{-1}. Any excess of the minimum for maintenance, 10 L water (mL O_2)$^{-1}$, is available for growth and reproduction, but if there is less than the minimum, there must be sources of food other than phytoplankton or it implies starvation.

Another interpretation of the F/R-ratio is related to the oxygen uptake. For example, the reciprocal of $F/R = 22.9$ L of water per mL of O_2, i.e., $R/F = 0.044$ mL O_2 (L water)$^{-1}$ may be compared to the actual O_2 content of water, which, at saturation, amounts to 6.3 mL O_2 (L water)$^{-1}$. The ratio, $EE = 0.044/6.3 = 0.007 = 0.7\%$ is denoted by the oxygen extraction efficiency. It is low for true filter feeders because the diffusive uptake readily provides the necessary oxygen, but it will increase if the flow is restricted by something, e.g., the reduced valve gape of mussels that occurs during periods of food depletion. The high value $EE = 1/(4.1 \times 6.3) = 0.039 = 3.9\%$ of *Verongia gigante* signifies an atypically high O_2 uptake, which is demanded by the symbiotic bacterial community within the sponge.

Finally, the metabolic cost of growth (R_g/G) that enters the bioenergetic growth model in the coefficient a_0 and represents the cost of synthesizing new biomass, which is mainly macromolecular synthesis, is generally low (8% to 23%, Table 2) but is very high (139%) for the sponge *Halichondria panicea* [11]. The latter high value is probably due to the early evolutionary simple structure that is mainly composed of the choanocyte pumps so that energy that is used for maintenance only constitutes a small fraction of the energy that is required for growth.

10. Conclusions

For the filter-feeding invertebrates that are examined herein, the results in Table 1 show that the theory of $b_1 \approx b_2$ appears to be approximately satisfied. The values of these b-exponents range from 0.66 to 1.0, implying size-specific growth rates of $\mu \approx W^{-0.34}$ to $W^0 =$ constant, i.e., from a power function growth that is decreasing with increasing size to exponential growth. Exponential growth is a feature of modularity, which among filter feeders only applies to some bryozoans and sponges. Here, the filtration rate of a bryozoan colony, for example, increases in proportion to its size, hence its number of individual zooids each have the same filtration rate. Similarly, the filtration rate and the growth increase in proportion to the size of a larger sponge if it is composed mostly of fully grown aquiferous units (modules) of an equal filtration rate. In addition to modularity, the growth may be near exponential in the early ontogenetic stages of filter-feeding invertebrates, such as mussels.

There seems to be no indication that b-exponents should take a suggested universal value of 3/4 [16], which is a trend that seems to appear when pooling data from a large number of organisms covering a large span of sizes [59,60]. The magnitude of size-specific growth rates range from less than 1% d^{-1} to more than 100% d^{-1} (Table 2), tending to be high for small organisms or those in the early stages of growth and decreasing with size.

Author Contributions: Conceptualization, writing—review and editing, P.S.L. and H.U.R. All authors have read and agreed to the published version of the manuscript.

Funding: This research received no external funding.

Institutional Review Board Statement: Not applicable.

Informed Consent Statement: Not applicable.

Data Availability Statement: Not applicable.

Conflicts of Interest: The authors declare no conflict of interest.

References

1. Jørgensen, C.B. *Biology of Suspension Feeding*; Pergamon Press: Oxford, UK, 1966; p. 358.
2. Wildish, D.; Kristmanson, D. *Benthic Suspension Feeders and Flow*; Cambridge University Press: Cambridge, UK, 1997; pp. 1–409.
3. Riisgård, H.U.; Larsen, P.S. Particle-capture mechanisms in marine suspension-feeding invertebrates. *Mar. Ecol. Prog. Ser.* **2010**, *418*, 255–293. [CrossRef]
4. Riisgård, H.U.; Lundgreen, K.; Larsen, P.S. Potential for production of 'mini-mussels' in Great Belt (Denmark) evaluated on basis of actual and modeled growth of young mussels *Mytilus edulis*. *Aquac. Int.* **2014**, *22*, 859–885. [CrossRef]
5. Jørgensen, C.B. *Bivalve Filter Feeding: Hydrodynamics, Bioenergetics, Physiology and Ecology*; Olsen and Olsen: Fredensborg, Denmark, 1990.
6. Riisgård, H.U. Filter-feeding mechanisms in crustaceans. In *Life Styles and Feeding Biological, The Natural History of the Crustacea*; Oxford University Press: New York, NY, USA, 2014; Volume 2, pp. 418–463.
7. Hamburger, K.; Møhlenberg, F.; Randløv, A.; Riisgård, H.U. Size, oxygen consumption and growth in the mussel *Mytilus edulis*. *Mar. Biol.* **1983**, *75*, 303–306. [CrossRef]
8. Clausen, I.; Riisgård, H.U. Growth, filtration and respiration in the mussel *Mytilus edulis*: No regulation of the filter-pump to nutritional needs. *Mar. Ecol. Prog. Ser.* **1996**, *141*, 37–45. [CrossRef]
9. Riisgård, H.U.; Lundgreen, K.; Larsen, P.S. Field data and growth model for mussels *Mytilus edulis* in Danish waters. *Mar. Biol. Res.* **2012**, *8*, 683–700. [CrossRef]
10. Riisgård, H.U.; Larsen, P.S. Filtration rates and scaling in demosponges. *J. Mar. Sci. Eng.* **2022**, *10*, 643. [CrossRef]
11. Thomassen, S.; Riisgård, H.U. Growth and energetics of the sponge *Halichondria panicea*. *Mar. Ecol. Prog. Ser.* **1995**, *128*, 239–246. [CrossRef]
12. Møller, L.F.; Riisgård, H.U. Feeding, bioenergetics and growth in the common jellyfish *Aurelia aurita* and two hydromedusae, *Sarsia tubulosa* and *Aequorea vitrina*. *Mar. Ecol. Prog. Ser.* **2007**, *346*, 167–177. [CrossRef]
13. Frandsen, K.; Riisgård, H.U. Size dependent respiration and growth of jellyfish (*Aurelia aurita*). *Sarsia* **1997**, *82*, 307–312. [CrossRef]
14. Riisgård, H. Suspension feeding in the polychaete *Nereis diversicolor*. *Mar. Ecol. Prog. Ser.* **1991**, *70*, 29–37. [CrossRef]

15. Nielsen, A.M.; Eriksen, N.T.; Iversen, J.J.L.; Riisgård, H.U. Feeding, growth and respiration in the polychaetes *Nereis diversicolor* (facultative filter-feeder) and *N. virens* (omnivorous)—A comparative study. *Mar. Ecol. Prog. Ser.* **1995**, *125*, 149–158. [CrossRef]
16. Kiørboe, T.; Hirst, A.G. Shifts in mass scaling of respiration, feeding, and growth rates across life-form. Transitions in Marine Pelagic Organisms. *Am. Nat.* **2014**, *183*, E118–E130. [CrossRef] [PubMed]
17. Petersen, J.; Riisgård, H. Filtration capacity of the ascidian *Ciona intestinalis* and its grazing impact in a shallow fjord. *Mar. Ecol. Prog. Ser.* **1992**, *88*, 9–17. [CrossRef]
18. Shumway, S.E. Respiration, pumping activity and heart rate in *Ciona intestinalis* exposed to fluctuating salinities. *Mar. Biol.* **1978**, *48*, 235–242. [CrossRef]
19. Riisgård, H.U.; Larsen, P.S. Actual and model-predicted growth of sponges—with a bioenergetic comparison to other filter-feeders. *J. Mar. Sci. Eng.* **2022**, *10*, 607. [CrossRef]
20. Lüskow, F.; Riisgård, H.U. Population predation impact of jellyfish (*Aurelia aurita*) controls the maximum umbrella size and somatic degrowth in temperate Danish waters (Kertinge Nor and Mariager Fjord). *Vie Et Milieu* **2016**, *66*, 233–243.
21. Ishii, H.; Båmstedt, U. Food regulation of growth and maturation in a natural population of *Aurelia aurita* (L.). *J. Plankton Res.* **1998**, *20*, 805–816. [CrossRef]
22. Hermansen, P.; Larsen, P.S.; Riisgård, H.U. Colony growth rate of encrusting bryozoans (*Electra pilosa* and *Celleporella hyalina*): Importance of algal concentration and water flow. *J. Exp. Mar. Biol. Ecol.* **2001**, *263*, 1–23. [CrossRef]
23. Key, M.M. Estimating colony age from colony size in encrusting cheilostomes. In *Bryozoan Studies 2019, Proceedings of the eighteenth International Bryozoology Association Conference, Liberec, Czech Republic, 16–21 June 2019*; Czech Geological Survey: Prague, Czech Republic, 2020.
24. Vedel, A.; Riisgård, H.U. Filter-feeding in the polychaete *Nereis diversicolor*: Growth and bioenergetics. *Mar. Ecol. Prog. Ser.* **1993**, *100*, 145–152. [CrossRef]
25. Giangrande, A.; Petraroli, A. Observations on reproduction and growth of *Sabella spallanzanii* (Polychaeta, Sabellidae) in the Mediterranean Sea. In *Actes dé la 4ème Conférence internationale des Polychètes*; Dauvin, J.-C., Laubier, L., Reish, D.J., Eds.; Mémoires du Muséum National d'Histoire Naturelle: Paris, France, 1994; Volume 162, pp. 51–56. ISBN 2-85653-214-4.
26. Lin, K.Y.; Sastri, A.R.; Gong, G.C.; Hsieh, C.H. Copepod community growth rates in relation to body size, temperature, and food availability in the East China Sea: A test of metabolic theory of ecology. *Biogeosciences* **2013**, *10*, 1877–1892. [CrossRef]
27. Petersen, J.K.; Schou, O.; Thor, P. Growth and energetics in the ascidian *Ciona intestinalis*. *Mar. Ecol. Prog Ser.* **1995**, *120*, 175–184. [CrossRef]
28. Petersen, J.K.; Schou, O.; Thor, P. In situ growth of the ascidian *Ciona intestinalis* (L.) and the blue mussel *Mytilus edulis* in an eelgrass meadow. *J. Exp. Mar. Biol. Ecol.* **1997**, *218*, 1–11. [CrossRef]
29. Nielsen, C. Six major steps in animal evolution: Are we dervied sponge larvae? *Evol. Dev.* **2008**, *10*, 241–257. [CrossRef] [PubMed]
30. Larsen, P.S.; Riisgård, H.U. The sponge pump. *J. Theor. Biol.* **1994**, *168*, 53–63. [CrossRef]
31. De Goeij, J.M.; van den Berg, H.; van Oostveen, M.M.; Epping, E.H.; Van Duyl, F.C. Major bulk dissolved organic carbon (DOC) removal by encrusting coral reef cavity sponges. *Mar. Ecol. Prog. Ser.* **2008**, *357*, 139–151. [CrossRef]
32. Leys, S.P.; Yahel, G.; Reidenbach, M.A.; Tunnicliffe, V.; Shavit, U.; Reiswig, H.M. The sponge pump: The role of current induced flow in the design of the sponge body plan. *PLoS ONE* **2011**, *6*, e27787. [CrossRef]
33. Riisgård, H.U.; Kumala, L.; Charitonidou, K. Using the F/R-ratio for an evaluation of the ability of the demosponge Halichondria panicea to nourish solely on phytoplankton versus free-living bacteria in the sea. *Mar. Biol. Res.* **2016**, *12*, 907–916. [CrossRef]
34. Reiswig, H.M. In situ pumping activities of tropical Demospongiae. *Mar. Biol.* **1971**, *9*, 38–50. [CrossRef]
35. Lüskow, F.; Riisgård, H.U.; Solovyeva, V.; Brewer, J.R. Seasonal changes in bacteria and phytoplankton biomass control the condition index of the demosponge *Halichondria panicea* in temperate Danish waters. *Mar. Ecol. Prog. Ser.* **2019**, *608*, 119–132. [CrossRef]
36. Bergquist, P.R. *Sponges*; University of California Press: Berkeley, CA, USA, 1978.
37. Reiswig, H.M. Water transport, respiration and energetics of three tropical marine sponges. *J. Exp. Mar. Biol. Ecol.* **1974**, *14*, 231–249. [CrossRef]
38. Larsen, P.S.; Riisgård, H.U. Pumping rate and size of demosponges—Towards an understanding using modeling. *J. Mar. Sci. Eng.* **2021**, *9*, 1308. [CrossRef]
39. Kumala, L.; Riisgård, H.U.; Canfield, D.E. Osculum dynamics and filtration activity studied in small single-osculum explants of the demosponge *Halichondria panicea*. *Mar. Ecol. Prog. Ser.* **2017**, *572*, 117–128. [CrossRef]
40. Goldstein, J.; Riisgård, H.U.; Larsen, P.S. Exhalant jet speed of single-osculum explants of the demosponge *Halichondria panicea* and basic properties of the sponge-pump. *J. Exp. Mar. Biol. Ecol.* **2019**, *511*, 82–90. [CrossRef]
41. Fry, W.G. The sponge as a population: A biometric approach. *Symp. Zool. Soc. Lond.* **1970**, *25*, 135–162.
42. Ereskovskii, A.V. Problems of coloniality, modularity, and individuality in sponges and special features of their mor-phogeneses during growth and asexual reproduction. *Russ. J. Mar. Biol.* **2003**, *29*, 46–56. [CrossRef]
43. McMurray, S.E.; Blum, J.E.; Pawlik, J.R. Redwood of the reef: Growth and age of the giant barrel sponge *Xestospongia muta* in the Floridas Keys. *Mar. Biol.* **2008**, *155*, 159–171. [CrossRef]
44. McMurray, S.E.; Pawlik, J.R.; Finelli, C.M. Trait-mediated ecosystem impacts: How morphology and size affect pumping rates of the Caribbean giant barrel sponge. *Aquat. Biol.* **2014**, *23*, 1–13. [CrossRef]

45. Riisgård, H.U.; Larsen, P.S. Bioenergetic model and specific growth rate of jellyfish *Aurelia* spp. *Mar. Ecol. Prog. Ser.* **2022**, *688*, 49–56. [CrossRef]
46. Riisgård, H.U.; Madsen, C.V. Clearance rates of ephyrae and small medusae of the common jellyfish *Aurelia aurita* offered different types of prey. *J. Sea Res.* **2011**, *65*, 51–57. [CrossRef]
47. Xixing, L.; Xueming, Y.; Jianghu, M. *Biology of Marine-Fouling Bryozoans in the Coastal Waters of China*; Science Press: Beijing, China, 2001.
48. Kiørboe, T. Small-scale turbulence, marine snow formation, and planktivorous feeding. *Sci. Mar.* **1997**, *61*, 141–158.
49. Kimmerer, W.J.; Ignoffo, T.R.; Slaughter, A.M.; Gould, A.L. Food-limited reproduction and growth of three copepod species in the low-salinity zone of the San Francisco Estuary. *J. Plankton Res.* **2014**, *36*, 722–735. [CrossRef]
50. Riisgård, H.U.; Funch, P.; Larsen, P.S. The mussel filter-pump—present understanding, with a re-examination of gill preparations. of ciliary structure and function. *Acta Zool.* **2015**, *96*, 273–282. [CrossRef]
51. Larsen, P.S.; Lundgreen, K.; Riisgård, H.U. Bioenergetic model predictions of actual growth and allometric transitions during ontogeny of juvenile blue mussels *Mytilus edulis*. In *Mussels: Ecology, Life Habits and Control*; Nowak, J., Kozlowski, M., Eds.; Nova Science Publishers: New York, NY, USA, 2013; pp. 101–122.
52. Larsen, P.S.; Lüskow, F.; Riisgård, H.U. Too much food causes reduced growth of blue mussels (*Mytilus edulis*)—test of hypothesis and new 'high chl *a* BEG-model'. *J. Mar. Syst.* **2018**, *180*, 299–306. [CrossRef]
53. Larsen, P.S.; Filgueira, R.; Riisgård, H.U. Actual growth of mussels *Mytilus edulis* in field studies compared to predictions using DEB, BEG and SFG models. *J. Sea Res.* **2014**, *88*, 100–108. [CrossRef]
54. Jørgensen, C.B.; Kiørboe, T.; Møhlenberg, F.; Riisgård, H.U. Ciliary and mucus-net filter feeding, with special reference to fluid mechanical characteristics. *Mar. Ecol. Prog. Ser.* **1984**, *15*, 283–292. [CrossRef]
55. Yamaguchi, M. Growth and reproductive cycles of marine fouling ascidians *Ciona intestinalis, Styela plicata, Botrylloides violaceus,* and *leptoclinum mitsukurii* at Abu ratsubo-Moroiso inlet (Central Japan). *Mar. Biol.* **1975**, *29*, 253–259. [CrossRef]
56. Dybern, B.I. The life cycle of *Ciona intestinalis* (L.) f. *typica* in relation to the environmental temperature. *Oikos* **1965**, *16*, 109–131. [CrossRef]
57. Collin, S.; Johnson, L.E. Invasive species contribute to biotic resistance: Negative effect of caprellid amphipods on an invasive tunicate. *Biol. Invasions* **2014**, *16*, 2209–2219. [CrossRef]
58. Riisgård, H.U.; Larsen, P.S. Comparative ecophysiology of active zoobenthic filter feeding, essence of current knowledge. *J. Sea Res.* **2000**, *44*, 169–193. [CrossRef]
59. Fenchel, T. Ecology–potentials and limitations. In *Excellence in Ecology*; Ecology Institute: Oldendorf/Luhe, Germany, 1987.
60. Riisgård, H.U. No foundation of a '3/4 power scaling law' for respiration in biology. *Ecol. Lett.* **1998**, *1*, 71–73. [CrossRef]

MDPI

St. Alban-Anlage 66

4052 Basel

Switzerland

Tel. +41 61 683 77 34

Fax +41 61 302 89 18

www.mdpi.com

Journal of Marine Science and Engineering Editorial Office

E-mail: jmse@mdpi.com

www.mdpi.com/journal/jmse